Math Standards in Action

Amy Seely Flint, M.Ed. &
Mary Riordan Karlsson, M.A.

Teacher Created Materials

Teacher Created Materials, Inc.

Cover Design by Darlene Spivak

Made in U.S.A.

ISBN 1-55734-886-3

Order Number TCM 886

Table of Contents

Introduction

Math Standards in Action discusses possibilities for implementing a mathematically powerful program in the classroom. Chapters highlight such issues as a constructivist perspective on learning, National Council of Teachers of Mathematics goals and curriculum standards, the classroom climate, integrated learning, and alternative assessment practices. Throughout the chapters, examples are provided to support the ideas and concepts of an empowering mathematics program. Sample forms are also included to help you begin thinking about ways in which mathematical literacy may be developed in your classroom.

As you read through the chapters, we hope you will begin to understand the complexities and dynamic nature of mathematics. Within a mathematically powerful program, students are encouraged to view mathematics as a process with many different pathways to understanding. The pathways link knowledge learned in the classroom to everyday experiences. *Math Standards in Action* offers you an opportunity to reflect on the goals and standards envisioned by the National Council of Teachers of Mathematics as we prepare students for the twenty-first century.

A New Focus

Races vs. Journeys

A timer buzzes loudly and sharply. Groans emanate from inside the classroom. The tension is thick as the teacher, who has been monitoring the students, yells, "Pencils down." Students glance around nervously. How many problems did they finish? What about their neighbors? Did they answer more? The teacher begins calling out answers at a rapid pace, "24, 32, 60, 54, 18," etc., until the last question is answered. Students are circling those they missed. Numbers are added up and recorded at the top of the page. The papers are passed forward. The teacher, noticing the clock, says, "Okay, now take out your social studies book and turn to page 48."

Numbers are added up and recorded at the top of the page. The papers are passed forward.

What images do you draw from this scenario? If asked to describe what the students are feeling and experiencing, what would you say? What knowledge are they learning about mathematical concepts and application? What is the teacher experiencing in this event? How much information is she/he gaining from the numbers reported at the top of each page? How can these numbers be used in evaluating students' understandings?

Mathematics in the classroom is continuing to undergo reform. The past number of years have afforded the educational community an opportunity to rethink how mathematics is being addressed in the classroom. The *Curriculum and Evaluation Standards for School Mathematics* (National Council of Teachers of Mathematics, 1989) was published in response to concerns about how mathematics was being addressed in the classroom setting. As teachers and educators become more familiar and knowledgeable about the standards and goals proposed in the document, classroom settings are beginning to reflect new pathways to understanding mathematical concepts and processes. Students are experiencing math in relevant and purposeful ways. There are opportunities to investigate mathematical concepts. Problems often have more than one solution. And, in fact, divergent solutions are encouraged. These changes in mathematics education are similar to those occurring in the language arts reform movements. Students are participating in a curriculum designed to encourage critical thinking and problem-solving approaches. For many teachers and students, a different scenario of how mathematics is approached in the classroom setting is being enacted.

Mathematics in the classroom is continuing to undergo reform.

A timer buzzes loudly and sharply. Groans emanate from the classroom. The tension is thick as the teacher begins to write the final number on the board. Students glance nervously around the room. What number will be recorded? As the teacher writes the number, students begin checking their work to see if they have reached the same conclusion. A group of students working collaboratively have not reached the same conclusion. Hands are raised quickly, and these students are provided an opportunity to offer an explanation of their thinking processes in solving the problem. The other students and the teacher listen intently to the explanation. Some agree. Others have differing ideas. Discussions ensue. The classroom is filled with excitement and energy. There are some students who have referenced a local trade paper and a farmer's almanac to locate more information for the problem. Further discussion occurs, and, finally, students reach a consensus. The math period has blended with reading and social studies, and the teacher, noticing the clock, says "Okay, sorry we have to end this discussion, but it is time for lunch."

What images did you draw from this second scenario? If asked to describe what the students are feeling and experiencing, what would you say? What knowledge are they learning about mathematical concepts and application? What about the teacher? What is she/he experiencing? How much information is she/he gaining from the students' discussions? Can this information be used in making curricular decisions and guiding assessment practices?

In comparing the two scenarios, it is possible to witness the change in mathematics instruction as it occurs in classroom settings? The first scenario required students to complete a very decontextualized task, a page of basic multiplication facts. There was no discussion about the task; students were merely to answer and move on. Students viewed math as abstract computations. It held little relevance to their lives then or for the future. As the teacher records the grades in the grade book, there is limited knowledge gained about the students as mathematicians, as problem solvers. Many students in this first scenario think of mathematics as a race to the finish line. Time is of utmost importance. It is more important to discover the finish line and cross it than it is to understand how one got there.

The second scenario creates a different picture about mathematics in the classroom. Students are participating in discovering solutions to a real-life problem. Other resources are consulted for possible ideas. Students are able to discuss their own pathways to the answer. Ideas are offered and shared. New understandings emerge. Math is viewed as an integrated component of the curriculum, merging with other disciplines. The teacher is provided opportunities to listen and reflect on the thinking patterns of the students. Students are perceived as complex problem solvers able to create possible solutions. The students in this scenario do not think of mathematics as a race but, rather, as a journey. Journeys provide opportunities to explore. Students focus on the process as conclusions are reached and solutions discovered.

Traditionally, mathematics was an abstract, decontextualized discipline where advanced concepts and applications were only for the specialized few.

New Focus and Purpose for Mathematics

The mathematics reform movement addressed the changing needs of society and, consequently, schools. With new technologies constantly reshaping the ways we conduct our lives, it became critical that schools and curriculums were realigned with the needs of society. The information age is definitely upon us. Computers, electronic mail, and the Internet are ways in which communication has changed in our society. These changes signaled a need for rethinking the approach to mathematics within the classroom setting. Traditionally, mathematics was an abstract, decontextualized discipline where advanced concepts and applications were only for the specialized few. Much has changed with the advancement of technology. Now students are "using mathematics in a world where calculators, computers, and other forms of technology are readily available; where mathematics as a field of knowledge is rapidly changing; and where mathematics is continually being applied to more fields of study" (Webb, 1993, p. 1). This shift in how mathematical concepts and applications may be used in future settings resulted in a change in the goals and purposes of mathematics education in the classroom.

The goals for mathematics education have moved from emphasizing abstract operations and computational skills to emphasizing contextualized, multidimensional problems students may be faced with in their own lives. The National Council of Teachers of Mathematics identified five goals in the *Curriculum and Evaluation Standards for School Mathematics* (1989) to embrace the changes. These goals constitute the underlying framework upon which curricular standards were constructed. The restructuring of goals and curricular standards focused both on the content of what was taught and the way it was taught. The overarching goals for mathematics education in the classroom are these:

- Students learn to value mathematics.
- Students become confident in their abilities to do mathematics.
- Students become mathematical problem solvers.
- Students learn to communicate mathematically.
- Students learn to reason mathematically.

Students develop math literacy when the purposes for participating in mathematical inquiry are relevant and meaningful to them.

Mathematical Literacy

With these broad goals as the base for developing a meaningful and purposeful mathematics education program, students will develop mathematical literacy. Mathematical literacy refers to the students' abilities to solve mathematical problems in a variety of contexts, for varying purposes, and with appropriate tools and techniques. Math tasks occur in many different situations. Students use mathematical concepts when sharing cookies with their friends, figuring out whose house is further away from the school, deciding on a pattern for an art project, wondering what the favorite food is in the classroom, and many other real-life situations. The contexts for mathematical inquiry are abundant. Students developing mathematical literacy take the opportunities to explore and investigate mathematical concepts as they occur in the differing contexts. Math is viewed as occurring everywhere, not just in math class or during math period.

The purposes for mathematical inquiry are closely connected to the contexts. Students develop math literacy when the purposes for participating in mathematical inquiry are relevant and meaningful to them. Some of the tasks students engage in reflect the complexities of life situations, while others are purely investigations of ideas (California Department of Education, 1992). In either situation, the purpose for engaging in the task must be meaningful for the student. Developing a sense of purpose for the math task requires that the tasks are rich and complex. They are open and accessible to all students and often have a multitude of possible solutions. Students are encouraged to decide their own pathways to the solutions. Purpose gives meaning and organization to students' thinking (California Department of Education, 1992).

Knowing the appropriate tools and techniques for solving mathematical problems is essential to developing math literacy. "Mathematical tools and techniques enable students to use their hands and eyes to experiment and explore relationships" (California Department of Education, 1992, p. 22). Manipulatives, calculators, and computers are tools that enable students to explore concepts and number patterns, analyze data rather than memorize a series of steps, and perform drills on paper. Determining the appropriate techniques to use in solving math problems requires that students have intellectual knowledge about the concepts, as well as problem-solving knowledge of ways to approach tasks and problems. Selecting the tools and techniques appropriate for the tasks is an important aspect of becoming mathematically literate.

Knowing the appropriate tools and techniques for solving mathematical problems is essential to developing math literacy.

The Role of Constructivist Thinking

Supporting the efforts to rethink the ways in which mathematics is being addressed in the classroom setting are teachers and educators who have a constructivist perspective on learning and mathematics education. Constructivists view learning as occurring in socially negotiated contexts where students have opportunities to share and discuss their own thinking and interpretations. Vygotsky (1978), a leading theorist in constructivism, discusses the role of social interaction as a key component to the learning process. It is through social interactions with others and the environment that children construct meanings and interpret the world around them. Vygotsky identified a zone in which learning occurs. Terming it the "zone of proximal development," he said,

> *It is the distance between the actual development level as determined by independent problem-solving and the level of potential development as determined through problem solving, adult guidance, or in collaboration with more capable peers (p. 86).*

The "zone of proximal development" enables the student to develop and gradually internalize new understandings. The social interaction among participants in the learning event enables those who are discovering the concepts to be guided by those who are familiar with and understand the concepts. A term many teachers are more familiar with is scaffolding. Similar to "zone of proximal development," scaffolding provides students with opportunities to try new ideas and to be supported by those in the classroom. Students introduced to the mathematical concept of place value may not fully comprehend the hundreds, tens, and ones places when working through problems. Interactions with others, where problems are solved in collaboration with more capable peers or adults, provides students with support in learning the new concept. The more capable peer offers ideas and suggestions on possible ways of understanding the concept. With this assistance, students are able to gradually develop an understanding of place value until they are confident and capable of solving place value problems on their own. The interaction between and among peers and the teacher is critical to the construction of meaning. Students share approaches, conjectures, strategies, difficulties, and results with others when opportunities to discuss present themselves. Through these interactions and others, the classroom becomes a socially constructed learning environment.

Interactions with others, where problems are solved in collaboration with more capable peers or adults, provides students with support in learning the new concept.

A constructivist perspective of mathematical inquiry and learning views development occurring not as a linear sequence of events but through situated actions and changes (Dyson, 1995). These actions occur in an environment that is socially constructed. The math problems students confront should be meaningful and purposeful to their own lives, not just ones that are at the next step or level of difficulty. Students excel at higher order thinking skills, and it is a myth that they should master lower order skills first (Resnick, 1987). Engaging in mathematical inquiry that makes sense provides students with a meaningful and relevant purpose. Mathematical knowledge is utilized in situations where it makes sense to discover new ideas about the world around them (California Department of Education, 1992).

Along with social interaction, constructivists believe students' experiences and background knowledge are critical to the learning process. They bring to learning events their own experiences that impact the contexts for the situations. Students come to math tasks with a wealth of knowledge and experiences. The uniqueness of these experiences contributes to the myriad of pathways of development that emerge. Even before starting formal schooling, young students have experiences with numbers, geometry, and solving problems. Mathematical inquiry should capitalize on these experiences, enabling students to make connections between concepts presented in the classroom and the world around them. The cultural and language diversities among students in many classrooms are also reflected in the experiences students bring to math problems and tasks. As students share their own interpretations, it becomes evident that there are many pathways to approaching the given problem. Celebrating the diversity of backgrounds and experiences enables teachers and students to appreciate the array of ideas, approaches, strategies, and solutions (California Department of Education, 1992).

Thinking about mathematical literacy with a constructivist perspective makes it possible to frame mathematical inquiry as occurring in meaningful, cohesive social contexts. Participating in socially situated contexts, students are provided with opportunities to increase knowledge of mathematical concepts and skills. They construct their own interpretations of the world based on experiences. Through these experiences, students discover, explore, investigate, and analyze mathematical ideas. Mathematical literacy, then, involves many facets and dimensions of the learning process, including contexts, purposes, and personal experiences.

New Perspectives Lead to New Questions

Recognizing the continuing advancement of technology and having an increased understanding of how young students learn mathematical concepts provides you with tools to begin thinking about your approach to mathematics education in your classroom. Is your math program similar to the first scenario, where the emphasis is on basic knowledge of rote skills, or does your program reflect a more integrated, constructivist view of mathematics? As you rethink how your students access and learn mathematical concepts in the classroom setting, you may have parents and others who have concerns about the direction of the math program. If you are not sending home worksheets of addition problems but, instead, asking students to find solutions to a problem involving their own personal collections, parents may wonder and question the value of your mathematics program. It will be important for you to address the questions and to assure them that their children are becoming mathematically literate.

Participating in socially situated contexts, students are provided with opportunities to increase knowledge of mathematical concepts and skills.

The focus of many of their questions may center around wondering if this is new math all over again with the absence of basic skills. In response to these questions and others, you can highlight the goals identified by the National Council of Teachers of Mathematics and how these goals focus on mathematical literacy and power for all students, not just those who have a natural disposition towards math. The concern over learning computation skills can be addressed as you inform parents that while you recognize their importance, a series of worksheets might not be the most effective method of teaching computation skills. Rather, it is important for students to understand what they are doing and why so that computation skills are utilized effectively in problem situations. Emphasizing an expanded definition of basic to include aspects of problem solving, tools, techniques, contexts, and mathematical thinking will encourage parents to value the ways in which their children are learning mathematical concepts and ideas.

The questions and concerns parents and others have about the new focus for mathematics education are abundant. There may be some resistance as you begin to implement a math program emphasizing problem solving, reasoning, communication, and confidence. Capitalizing on opportunities to reassure parents will prove to be beneficial. Sharing literature with parents on the new approaches to mathematics, such as the contents of this book or inviting parents to participate in an activity with their children during Back to School Night, among other opportunities, will help to put parents at ease about your mathematics program.

Concluding Remarks

Mathematics education in the classroom is changing. The new changes focus on not only what is being taught but also how it is being taught. Teachers are beginning to rethink their own mathematics programs as they come to understand and learn more about the ways in which students learn mathematical concepts and skills. Students are being viewed as active contributors to their knowledge and understandings. Background experiences play an important role in how students come to access math concepts and skills. Multiple pathways to solutions are encouraged as students construct their own interpretations and understandings. The National Council of Teachers of Mathematics has embedded these constructivist perspectives of learning and mathematics development into its curriculum standards and goals. As you read through the chapters in this resource book, we hope you will see aspects of your own math program, as well as encounter new ideas and possibilities. *Math Standards in Action* integrates the goals and standards proposed by the National Council of Teachers of Mathematics into the classroom setting.

***Math Standards in Action* integrates the goals and standards proposed by the National Council of Teachers of Mathematics into the classroom setting.**

Goals and Directions

NCTM Goals for Mathematics Education

"But why do I have to learn this? I'm never going to use it when I grow up. Does it really matter which train is going faster?" These sentiments have been heard many times in the past. Perhaps you have even had the same thoughts as a student when performing computations on some abstract math problem assigned by the teacher. As the teacher, you may have also experienced a sense of abstractness in teaching operations and computations to a classroom full of blank faces. The math lessons were not engaging or relevant to the students. Consequently, there was a lack of motivation from students and teachers to discover that mathematics was more, much more, than a series of computations and decontextualized word problems.

As the teacher, you may have also experienced a sense of abstractness in teaching operations and computations to a classroom full of blank faces.

Mentioned in the first chapter, the National Council of Teachers of Mathematics (1989), in reaction to the changes in the way mathematics is used in society and in reaction to the call for reform, set out to develop goals and a direction that will prepare students to be mathematically literate in the twenty-first century. As these goals were developed, it became clear that the direction for mathematics education in our country was to ensure that all students were literate and confident in mathematical processes and strategies.

Mathematical literacy would involve more than understanding basic operations and computation skills. Students would have to develop problem-solving techniques and higher-level thinking and reasoning skills to understand issues in a technologically advanced society. With this direction in mind, the NCTM identified five goals for students in mathematics education. The goals were meant to serve as a framework to begin thinking about how mathematics should be addressed in the classroom setting. The goals include: learn to value mathematics, become confident in doing math, become a mathematical problem solver, learn to communicate mathematically, and learn to reason mathematically. Aiming for these broad goals will provide students with direction to develop their mathematical power and knowledge. Mathematical power is defined as "a student's abilities to use a variety of mathematical methods effectively to solve non-routine problems" (National Council of Teachers of Mathematics, 1989, p. 5). The suggested goals provide diverse avenues to travel as students and teachers explore mathematical concepts and processes.

Valuing mathematics also involves understanding the connection between theoretical and practical issues in math.

Goal 1: Learning to Value Mathematics

"Learning to value mathematics" is the first goal of mathematics education in today's classrooms. How does a teacher instill value into the students? What is value? According to the *Curriculum and Evaluation Standards for School Mathematics* (NCTM, 1989), the term value is synonymous with appreciation. Students will learn to appreciate the role of mathematics in our society. For instance, math permeates and is embedded in all aspects of our daily life. We use math concepts and skills to cook, to drive, to purchase items, to enjoy sports, and to understand the stock market, among many other daily events. The goal, then, is to encourage students to recognize the important role math plays in conducting routine activities in our lives.

Valuing mathematics also involves understanding the connection between theoretical and practical issues in math. Theory and practice are closely connected and often entangled with each other. It is important for students to realize the connection between theory and application. Theory proposes the possibility, while application, or practice, determines whether it is viable and possible. Students come to understand this interaction in an historical sense by reflecting on how theory has infused itself into the daily workings of their lives.

> *"It is the intent of this goal—learning to value mathematics—to focus attention on the need for student awareness of the interaction between mathematics and the historical situations from which it has developed and the impact that interaction has on our culture and our lives" (NCTM, 1989, p. 6).*

Goal 2: Becoming Confident in One's Own Ability

This goal speaks directly to the students' own perceptions of themselves as mathematicians. In past years, many students developed anxiety about doing mathematical problems. They were not provided opportunities to explore and discover math in their own ways. Instead, the expectation was to just learn the steps involved in the operation—never mind the process. The product or answer was the important part. The lack of discussion about how to solve the problem and the limited opportunities to process the conceptual knowledge within the problems resulted in students feeling insecure and not confident about their abilities in doing math.

To address this sense of uncertainty from the students, the NCTM recognized that if students were to become mathematically literate and demonstrate math power, they would have to become confident in their abilities. To develop this confidence, students must be provided with a multitude of situations and problems with many pathways to discovering solutions. Math problems presented with diverse solutions enable students to believe in themselves and their approaches to the problems. Also, confidence builds as students come to view math as occurring in everyday situations. Students who are confident in their abilities are "inclined to use mathematics to make sense of situations that come up in their lives and to achieve their own purposes" (California Department of Education, 1992, p. 25). Through the numerous and various experiences, students begin to trust their own pathways and thinking processes (National Council of Teachers of Mathematics, 1989). With this confidence in their ability to solve mathematical problems and situations, students engage in more complex problems, discovering higher levels of mathematical thinking and reasoning.

Problem solving is at the heart of mathematical literacy and power.

Goal 3: Becoming Mathematical Problem Solvers

Problem solving is at the heart of mathematical literacy and power. There has been much discussion about transforming the mathematics curriculum into one that reflects a problem-solving approach. Those involved with mathematics education recognize the importance of this goal and have consistently encouraged teachers to provide more problems and situations for students. Yet, with all of this interest, it is still necessary to identify problem solving as a goal. NCTM urges all students to develop their abilities to solve problems in a number of different contexts.

Different contexts involve reconfiguring time devoted to mathematical inquiry and grouping students. It is important for students to be provided opportunities to work through problems involving different amounts of time. Some problems may take a short amount of time, ten to fifteen minutes or so; others involve longer periods, such as days or even weeks. The group configurations students work in also vary. Sometimes it is beneficial to work in a group, each member contributing to the collective solution. Other times, individual work is important. Encouraging students to solve a range of problems in different contexts and settings provides a sense that real-world mathematics is complex and varied.

Students engaged in problem solving utilize aspects of critical thinking, including analyzing a problem situation, making decisions, monitoring progress, and evaluating the completed solution (Szetela, 1993). These critical thinking skills are developed and explored as students offer possible solutions to the problem presented. The goal, then, of becoming mathematical problem solvers is a process whereby students utilize various critical thinking skills and strategies to support a reached conclusion.

Encouraging students to solve a range of problems in different contexts and settings provides a sense that real-world mathematics is complex and varied.

Goal 4: Learning to Communicate Mathematically

The fourth goal, learning to communicate mathematically, encourages students to express their understandings of mathematical concepts through various modes of communication. Included within these modes are written and verbal systems. As students come to explore and investigate mathematical concepts, it is important that they are able to communicate their understandings and processes. Learning to communicate mathematically involves students learning the signs, symbols, and terms of mathematics. Students are able to demonstrate their abilities to communicate mathematically by engaging in opportunities to read, write, draw, and discuss ideas and concepts in mathematical contexts. Through communication, it becomes possible for others to reflect upon the process and to offer new possibilities and insights into the thinking patterns.

Students communicate their thinking and understandings in a variety of ways. Included may be informal conversations and discussions, presentations, diagrams, graphs, tables, narratives, and symbols. "Communication helps to clarify a student's thinking, and feedback (formal or informal) can provide useful information for revision" (California Department of Education, 1992, p. 21). As students engage in the many modes of communication, their understandings become refined and focused.

Goal 5: Learning to Reason Mathematically

An important goal in developing students who possess mathematical literacy and power is the goal of learning to reason as a mathematician. Reasoning is a thinking process where logical conclusions and justification are emphasized. Students make conjectures, gather evidence, and build an argument to support the solution. Through the development of reasoning skills and strategies, students become autonomous learners. They are able to take control of their own successes and failures, justifying answers in a logically sound argument.

Reasoning skills, much like problem solving, permeate all aspects of the mathematics curriculum. For students, reasoning involves thinking about what makes sense for the problem or situation. For any problem or situation, students should be able to explain and justify their thinking. The explanation and justification processes are important to students as they develop mathematical power. Offering an appropriate solution is not adequate if the student cannot explain how the conclusion was reached. Mathematical reasoning does not develop in isolation. "It is a process that grows out of many experiences that convince children that mathematics makes sense" (National Council of Teachers of Mathematics, 1989, p. 31).

Reasoning skills, much like problem solving, permeate all aspects of the mathematics curriculum.

From Goals to Direction

The identified goals provide the direction teachers and students take in mathematics education. The direction of mathematics education has moved away from a behaviorist tradition, which focuses on what students do, to a constructivist tradition, where the focus is on how students construct knowledge and understandings (Battista & Larson, 1994). The goals have opened up mathematics, making understandings accessible from the many pathways students and teachers travel. Students explore, discover, think, reason, classify, analyze, plan, investigate, and formulate, among other thinking processes, in relevant and purposeful contexts. The direction, then, based on the five goals, is towards mathematical literacy and power, understanding what it means to know and do mathematics.

Knowing mathematics is, in fact, doing mathematics. It is an active process whereby the student gathers, discovers, and creates knowledge based on prior experiences and backgrounds. "This active process is different from mastering concepts and procedures" (National Council of Teachers of Mathematics, 1989, p. 7). Mastering concepts and procedures is valuable in a mathematics program as long as the concepts being mastered are relevant and useful to the students and the activity. For example, students learning basic multiplication facts find the process to be active when the knowledge gained relates to their own worlds and the information is meaningful to discovering relationships and connections to other areas. Multiplication encourages students to discover a more efficient method of adding numbers and quantities together. On the other hand, learning multiplication would not be an active process if the students were expected to memorize the facts and not move beyond to applying the knowledge in any real context for the students. The building of the active knowledge and the process of making decisions is doing mathematics.

The direction for mathematics education, then, appears to be one where students and teachers are active learners and thinkers. It is a process of inquiry. Constructing knowledge occurs in authentic contexts and relevant settings. Students are able to take diverse pathways in solving problems and discovering solutions. As active learners in the classroom, your students have opportunities to interact with each other, discussing and negotiating possible solutions. They are problem solvers, mathematicians.

The direction for mathematics education, then, appears to be one where students and teachers are active learners and thinkers.

An Empowered Mathematics Program

With mathematics education taking a new direction, one that is active and constructivist oriented, the math program in a classroom setting is moving from a static, teacher-directed program to one that is dynamic and empowering for students. In order for your students to be confident in their abilities and learn to value mathematics in their lives, a mathematics program should incorporate a few essential characteristics. Identified in the state framework for California (California Department of Education, 1992), these characteristics include, but are not limited to, all students should participate to the fullest extent, teachers are facilitators to knowledge, rather than owners of knowledge, all students have access to and regularly use a wide variety of materials and tools to solve problems, discussions and dialogues are emphasized as students work together, assessment practices are integrated with instruction, and all students have opportunities to reflect on their thinking, orally and in writing (California Department of Education, 1992). While these characteristics were defined in a state framework, they are generalizable to states and districts throughout the country. The characteristics offer suggestions of ways in which your mathematics program in the classroom can move from static and teacher directed to one of inquiry and empowerment.

The first characteristic, all students should participate to the fullest extent, highlights the importance of providing opportunities for students to investigate problems beyond the standard expectation. Your students should explore a vast array of problems, not just those limited to the prescribed grade level. Also, your students come to mathematics with diverse cultural, linguistic, and economic backgrounds and experiences. Honoring the diversity and using students' own experiences as the context encourages your students to be more active and engaged in the mathematics program.

Second, teachers as facilitators to knowledge, rather than owners of knowledge, is aligned with the idea of moving from the behavioristic mode of teaching to a constructivist mode. As you develop your mathematics program, opportunities to support and encourage your students will be abundant. The environment will be rich for your students to investigate problems and possible solutions. You will provide your students with probing questions as you discover their thinking patterns and ways of constructing understandings.

The third characteristic to developing an empowering mathematics program involves materials and tools. All of the students in your classroom should have access to a wide variety of resources. Manipulatives and technology are tools to explore mathematical ideas and concepts. With the resources available, your students will have to make decisions about which tool is most appropriate for the problems. Decisions such as these encourage the development of problem-solving techniques and are empowering for the students.

As you create lessons and activities for your students to learn mathematical concepts, the assessment practices should also reflect this learning.

Fourth, emphasizing dialogues and discussions in a mathematics program speaks directly to the idea that learning is a socially constructed event. Your students have much to offer and share. Through the interactions, knowledge is constructed, negotiated, and reinterpreted. Providing opportunities for students to dialogue and discuss mathematical concepts is important to exploring the diverse pathways students take to construct knowledge and understandings.

The fifth characteristic, integrating assessment and instruction, is critical to developing an empowering, active mathematics program. As you create lessons and activities for your students to learn mathematical concepts, the assessment practices should also reflect this learning. Aligning the instruction with assessment enables your students to see the value in mathematics and to know that what is taught will be assessed.

Finally, the sixth characteristic identified in developing an active, student-empowered program is reflection. You and your students may reflect in writing or through discussions. Reflection enables your students to clarify their thoughts and communicate their ideas and results. As you encourage students to reflect on their own thinking processes, you will be able to also gain an understanding of how your students approach and solve problems.

The six characteristics are possible suggestions for moving the program towards an active, student-empowered math program. Empowering your students provides them with a voice in the decision-making process. With a voice, they are more motivated and interested in the learning process. The increased motivation will encourage your students to develop a better understanding of mathematical concepts and ideas. They will become problem solvers and utilize their reasoning skills as solutions are investigated and analyzed.

Concluding Remarks

The mathematics program in many classrooms throughout the country is beginning to reflect the goals and direction emphasized by the National Council of Teachers of Mathematics. These goals highlight the importance of students being engaged and motivated to discover and investigate mathematical concepts and ideas. The goals also encourage teachers and students to rethink the ways in which mathematics is being taught in the classroom setting. Students should have opportunities to engage in mathematical inquiry through problem-solving and reasoning skills. As they participate in exploring mathematical concepts, students are communicating their understandings in a multitude of ways. Reading, discussing, writing, and drawing math are possible ways to demonstrate knowledge about mathematics and the world around them. The goals enable students and teachers to view mathematics in relevant and meaningful contexts.

Students should have opportunities to engage in mathematical inquiry through problem-solving and reasoning skills.

The direction mathematics education is moving towards, based on the identified goals, is one where students are viewed as active and contributing members to the learning process. Students are also provided opportunities to reach beyond the standard expectations and explore math concepts at a deeper level. Through the goals, direction, and characteristics of an empowering mathematics program, students will have opportunities to discover, investigate, conjecture about, wonder about, and explore the many facets of math and all that math has to offer.

The Classroom Climate

Opening Up Classroom Doors

Learning is a multifaceted, complex process involving students, teachers, tasks, and environment. Bringing all of these together in the learning process is challenging. Teachers must structure the curriculum to engage students in meaningful and relevant tasks. Students must participate and contribute to the tasks and environment. In teaching mathematical ideas and concepts, it is important to recognize factors that may contribute to the effectiveness of the learning process such as, motivation to learn, classroom environment, social interaction, and cooperative learning strategies. Each of these factors, discussed in detail, illuminates the constructivist perspective of learning and teaching. The physical and emotional elements in a classroom work together to make learning come alive and contribute to the intellectual, social, and emotional development of the students.

Learning is a multifaceted, complex process involving students, teachers, tasks, and environment.

Upon entering an elementary classroom, a child will look around and see evidence of math literacy everywhere—the clock on the wall, number charts, graphs, calendars, and schedules. This entrance into such a learning environment can either be exciting or daunting. The motivation level may rise or fall, depending on a few variables, specifically, past experiences with math, the pressure to find the right answer quickly, and the design of the curriculum.

Motivation to Learn

Many factors contribute to a student's motivation to learn. These factors can be personal, affective, cognitive, and social. Individuals are predisposed by personal factors and their personality characteristics while engaged in the learning process. The personal factors are those that a student brings into the classroom, such as purposes, the plan which guides the student to accomplish his/her goal; goals, an end or objective the learner sets in his/her mind; self-worth, the combination of self-esteem and self-confidence a learner possesses; self-regulation, which suggests that a person is able to complete tasks on his or her own, drawing on internalized competencies and strategies (Spaulding, 1992); self-determination, having the control over tasks to be completed; and, finally, family values and beliefs which are woven throughout these many influences, both directly and indirectly. These factors can originate from family traditions and cultural beliefs in the home environment or in early experiences in the school. Other personal factors that influence a learner in many different ways are called affective influences, those influences which directly affect the learning outcome. Affective influences involve an interplay between the students' own personal factors and outside sources. These influences consist of motivation, attitude, stance, response, and responsibility and can sway a learner one way or the other (Karlsson, 1996).

Motivation to investigate, discover, reason, problem solve, and compute is closely connected to opportunities to experience a sense of enjoyment while engaged in the activity.

Now that personal and affective factors have been identified as influencing students in the learning process, it is important to understand those influences within a student's head, known as cognitive influences. These require thinking skills, as opposed to the affective influences which rely more heavily on feelings and emotions. Cognitive influences definitely play a prominent role in the learning process, specifically in background knowledge, creating images (or imagery), self-monitoring skills, and word analysis skills (which are needed to read word problems). Many of these influences constantly interact with the personal and affective influences to help a learner comprehend and understand the subject matter.

Motivation to investigate, discover, reason, problem solve, and compute is closely connected to opportunities to experience a sense of enjoyment while engaged in the activity. To truly experience the intrinsic rewards of learning, students must achieve a balance between the difficulty of the task and their own ability. If the challenge of the task is too high, the learner will experience anxiety, and, conversely, if the challenge is too low, boredom will result.

There are two types of motivation, extrinsic (external) and intrinsic (internal), both of which are formative parts of the learning process. Extrinsic motivation is something that engages students in an activity when it is clear that by participating in the activity, a positive experience and desired outcomes are likely to result. Examples of extrinsic motivation are good grades, stickers, or special privileges. The extrinsic factors that influence motivation include the setting, the difficulty of the material, peer pressure or peer interaction, the goal of the assignment, the teacher's instructional strategy, and the amount of time given to complete the activity and communicate ideas and concepts. Most of these factors are not in the student's control; however, many of the intrinsic factors can be controlled by the students, such as their own personal goals or purposes, their self-worth and self-confidence, their attitudes, and their responses to the activity. Intrinsic motivation is characterized by a desire to engage in an activity because doing so brings personal satisfaction, regardless of potential extrinsic outcomes. An example of intrinsic motivation is personal satisfaction in discovering a solution to a problem. Both the extrinsic and intrinsic factors promote motivation to learn and the intention to participate in any activity.

Historically, the importance of the physical environment in learning and literacy development has been emphasized by many theorists and philosophers.

Our classrooms need to motivate students to use math to reach beyond the classrooms, schools, and communities into all aspects of their lives. The teacher and the classroom learning environment play a vital role in the development of young minds by stimulating those internal motivations and creating the interest and desire to learn within the classroom. The external motivating factors within the classroom include learning opportunities, setting, language interaction, and cooperative learning strategies. All of these factors are integrated within the physical aspects of the room to create the classroom learning environment which provides the interest and motivation to learn. While these factors overlap, interrelate, and interact with one another, intention, desire, and interest are included in all of these areas and are the result of the combination of these various influences.

The Instructional Setting

Historically, the importance of the physical environment in learning and literacy development has been emphasized by many theorists and philosophers. According to Piaget, children acquire knowledge by interacting with the world or the environment (Piaget & Inhelder, 1969). In the classroom learning environment, children can assimilate new experiences into what they already know. Learning takes place as the child interacts with peers and adults in social settings and conducive environments (Vygotsky, 1978). Based on these theories, the classroom should be designed to provide optimum development in mathematical literacy through problem-solving activities across the curriculum that include reading, writing, speaking, listening, experimenting, computing, and drawing. These problems allow students to discover and explore the world around them while using mathematical ideas and concepts.

The physical setup of the room has evolved over the years to reflect the changing theories of learning. Rarely will you see a classroom with the desks in rows and the teacher's desk in the front by the blackboard. Rather, in a constructivist classroom, the desks are set up in groups so the students can work together on various activities and assignments. You may see a bin of materials with scissors, markers, rulers, pencils, erasers, protractors, math manipulatives, etc., in the middle of the group for the students to share. The room may be set up in centers which emphasize one subject area, for example, a math center. Although these centers emphasize one particular area, the activities within each center integrate many subjects. For instance, in the math center there may be word problems requiring reading and writing, or in the science center, a seed-planting activity requiring reading and measuring.

External influences in the classroom, such as peer interaction and teacher interaction, often provide learners with a purpose and a goal for learning.

The setting, or social and physical surroundings, is important for many reasons, especially in a learner's motivation to learn. It must be compatible with the activity, or else the learner will not feel very comfortable. For instance, if a student is working out a math problem with pattern blocks in the math center and the rest of the class is singing a song with a record, the student who is trying to concentrate on math will feel uncomfortable. On the other hand, if the classroom is dull and boring and a learner wants to talk about the results from a math puzzle with his/her partners, the setting is not compatible with that activity and may deter the learner from initiating this important discussion.

External influences in the classroom, such as peer interaction and teacher interaction, often provide learners with a purpose and a goal for learning. These, then, combine with the instructional setting to influence the learner. The classroom environment should promote autonomy, which, in turn, will have a positive effect on the student's intrinsic motivation for learning. Thus, the social context is crucial in the development of internal motivations and positive attitudes towards learning.

Empowerment Through Interaction

Interactions occur on a daily basis in the home environment and in the classroom community, and it is through these interactions that "things act and react upon one another, and in so doing are reciprocally transformed" (Dewey, 1938, p. 42). Dewey also insisted "that all thought must refer to experience, that all valid knowledge and sound practical judgment must proceed from the analysis of experience" (p. 42). These interactions create experiences children use to learn new knowledge and build on prior knowledge. All ideas are thus connected to experiences and the environments in which they happen. Math is a socially mediated interactive process including the student, the task, a teacher, and the environment. Therefore, learning to think mathematically and value mathematics happens through social interaction and dialogue with others and is embedded in various math literacy activities within familiar environments.

Although previous conceptions of teaching and learning emphasized the teacher's role in transmitting knowledge, more recent theories have emphasized the social and interactional nature of learning (Turner & Paris, 1995). Teachers and students often develop an apprenticeship relationship where a more able companion guides, supports, and challenges another's understanding. Rogoff (1990) further elaborates on the idea of "zone of proximal development" with her concept of guided participation which involves children observing and participating with peers and more skilled members of their society in various problem-solving tasks. It is through the social relationship between the child and the other participant that learning occurs and new understandings and skills are acquired.

Math learning can be achieved through a strategy known as cooperative learning, which is more than just pushing desks together to form a small group.

Learning occurs within the social context when the student's experience involves social activity in the sense that he or she participates in interactions involving one or more persons. The importance of the social activities happening in the classroom learning environment is tremendous. There is an inexhaustible amount of mental and physical activity in the classroom from which each child can gain different knowledge in every social interaction. Math and reading are two of the most important content areas that promote social interaction activities within the classroom context. These occur within writing activities, science experiments, math games, and social studies projects, as well as music performances and art activities. Sometimes even physical fitness activities require math and reading skills.

Cooperative Learning

Math learning can be achieved through a strategy known as cooperative learning, which is more than just pushing desks together to form a small group. Rather, cooperative learning is a strategy that requires students to work cooperatively towards a common goal in small heterogeneous groups. Each student has an individual task to carry out without the direct supervision of the teacher. In essence, the students supervise each other and combine their efforts while working toward the accomplishment of instructional goals. Interdependent grouping encourages students to interact and problem solve in real-life situations.

Cooperative learning has two essential characteristics: a cooperative goal structure (positive interdependence) and individual accountability (Grisham & Molinelli, 1995). The goal structure guides the students and identifies the type of interdependence among the group and different ways the students will relate to each other. The three most common goal structures are cooperative, competitive, and individualistic. A cooperative goal structure exists when students believe they can reach their own goals if, and only if, the other students in their group can reach their goals. There is a good deal of social interaction in this type of goal structure. Conversely, a competitive goal structure exists when students believe they can reach their own goals if, and only if, the other students fail to reach their goals. Perhaps there may be some social interaction in this type of goal structure; however, it tends to be negative. An individual goal structure exists when a student achieves her/his goal on her/his own, unrelated to other students. Usually, there is limited social interaction in a situation with an individualistic goal structure.

Cooperative learning naturally fits into math literacy tasks by encouraging students to learn mathematics and solve problems jointly.

The second essential characteristic of cooperative learning is individual accountability; students are held accountable for their assigned tasks. Knowing that classmates are depending on them motivates students to participate enthusiastically to complete the task at hand. The expectations should be clear at the beginning of the task, and roles may be assigned to group members, such as supplier, reporter, recorder, encourager, etc. Alternative forms of assessment are common for cooperative activities since both the individual and the group are evaluated. Portfolios are a good tool for assessing the individual. Portfolios allow students to review their work and reflect on their growth and development as individuals and as team players. Other forms of assessment are needed to evaluate the group. Perhaps one of the more promising assessment alternatives involves the use of teacher or class-designed evaluation forms and rubrics for evaluating individual and group performance (Ryan, 1994). Thus, students can be evaluated by peers, as well as the teacher. More on assessment will be covered in chapter seven "Alternative Assessment Practices."

Cooperative learning naturally fits into math literacy tasks by encouraging students to learn mathematics and solve problems jointly. Some mathematical concepts and ideas may not seem as overwhelming if students are afforded the opportunity to discuss the concepts with classmates. Together, students can explore, discover, reflect, and clarify their thinking about mathematical ideas and situations. Through activities, such as counting with base ten blocks, making patterns with attribute blocks, or changing dollars into cents, students can relate their everyday language to mathematical language and symbols while they realize that discussing, reading, writing, and listening to mathematics are a vital part of learning and using mathematics.

There are many benefits of cooperative learning in the classroom. As a strategy, it tends to promote higher achievement rather than competitive or individualistic learning situations (Johnson, Maruyama, Johnson, Nelson, & Skon, 1981). As students share more opportunities to work together on assignments, they build a positive atmosphere within the classroom, replacing the natural urge to compete. Students are afforded a chance to be valued by their classmates, to feel good about themselves. In addition, this kind of structure in the classroom promotes many social outcomes, such as leadership ability, communication, and conflict management skills (Grisham & Molinelli, 1995).

When selecting the members for cooperative groups, be sure to create a mix of achievement, language, and social abilities.

Determining cooperative groups may be a difficult task in the beginning. It is important to know your students and to know guidelines for the selection of groups. When selecting the members for cooperative groups, be sure to create a mix of achievement, language, and social abilities. Three to five students per group is a good number, as far as group size is concerned. In cooperative learning situations, an uneven number is an asset during the decision process. Essentially, there can be no ties. Therefore, if you have twenty-eight students, you will create five groups, each consisting of five students, and one group of three. It is acceptable to have a few remaining students make up a smaller group.

The ideal cooperative learning situation should be thought of as a salad bowl as opposed to a melting pot. In a salad bowl concept, all individual learning types mix together. They create synergy while remaining individuals (Jasmine & Jasmine, 1995). In a melting pot situation, supposedly, individual traits melt together; thus, students lose their individuality. Cooperative learning holds the potential for both honoring and nurturing the diverse intelligences manifested within any classroom (Grisham & Molinelli, 1995).

Concluding Remarks

In sum, motivation to learn, the instructional setting, social interaction, and cooperative learning work together to achieve the goals set forth in chapter two "Goals and Directions." These classroom aspects encourage students to learn to value mathematics and become confident in their own abilities. The social interaction that takes place during mathematics instruction in the classroom and the opportunity to work in cooperative learning groups contributes to becoming a mathematical problem solver and learning to reason mathematically. The classroom is a very busy social arena where much learning takes place. Within this social context, there is a generous amount of language interaction between the teacher and students, meaningful math tasks for the students to engage in, and activities that provide opportunities for growth and development. These opportunities promote the communication of mathematical concepts and ideas, which leads to a greater understanding and appreciation for mathematics.

Curriculum Standards

Value Judgments

The *Curriculum and Evaluation Standards for School Mathematics* (NCTM, 1989) was devised to meet the mathematical needs of the students preparing for the twenty-first century. The intent of this document was to provide a broad framework to guide the restructuring movement in school mathematics into the next decade, based on these tenets: all students need to learn more (and often different) mathematical content than is contained in current programs and the currently prevalent methods of instruction in mathematics must be significantly revised (California Department of Education, 1992). This document also provides for grades kindergarten through twelve a set of standards defined as a "statement that can be used to judge the quality of a mathematics curriculum or of methods of evaluation" (California Department of Education, 1992, p. 78). Basically, a standard is a statement about what is valued, and these standards articulate the five overarching goals that were discussed in detail in chapter two "Goals and Directions."

It is important to know that these standards represent a broad perspective of mathematical literacy.

It is important to know that these standards represent a broad perspective of mathematical literacy. This chapter discusses the standards for an elementary mathematics curriculum. The first four standards deal with problem solving, communication, reasoning, and making connections, and these involve the process of doing mathematics. Standards five through fifteen involve the content standards.

Standard 1: Mathematics as Problem Solving

The standards suggest the study of mathematics should emphasize problem solving so that students can use problem-solving approaches to investigate and understand mathematical content, formulate problems from everyday and mathematical situations, develop and apply strategies to solve a wide variety of problems, verify and interpret results with respect to the original problems, and acquire confidence in using mathematics meaningfully.

The process of problem solving should be the primary focus of the mathematics curriculum; thus, it is a major goal of instruction and an essential element in all mathematical activities. Problem solving is a process that should permeate the entire math program and provide context and opportunities to learn concepts and skills.

Problem solving is a process that should permeate the entire math program and provide context and opportunities to learn concepts and skills.

Developing and applying strategies to solve problems is the major goal of problem-solving instruction. It is important to teach students there is more than one way of solving a problem without specific rules. Such strategies may include using trial and error, drawing diagrams, finding patterns, and role-playing a problem. These strategies encourage students to use higher-level thinking processes, such as analyzing, synthesizing, and inquiry. Additionally, through problem-solving activities, students communicate using mathematical concepts and terms and build confidence in their mathematical abilities.

A very important aspect of mathematical instruction, especially in solving problems, is having students formulate their own problems. Students can also rewrite problems that have already been solved, using new information and familiar names and settings. Another way for students to enhance their problem-solving abilities is through group projects which encourage students to work together to figure out answers and solutions. Samples of group project problems within interdisciplinary units are offered in chapter six "Linking Math to Our World."

Standard 2: Mathematics as Communication

Your mathematics curriculum should provide several opportunities for communication so students can "relate physical materials, pictures, and diagrams to mathematical ideas; reflect on and clarify their thinking about mathematical ideas and situations; relate their everyday language to mathematical language and symbols; realize that representing, discussing, reading, writing, and listening to mathematics are a vital part of learning and using mathematics" (NCTM, 1989, p. 26).

Speaking the language of mathematics in the classroom enables students to make mathematics a part of their lives and clarify their thinking. Through this communication, students make links among prior knowledge, new knowledge, intuitive notions, and the abstract language and symbolism of mathematics. When students are afforded the opportunities to see that one representation can be described in many situations, for example an equation, they begin to understand the power of mathematics.

Humans are social by nature; therefore, giving students the chance to talk about ideas and findings develops key communication skills such as representing, listening, talking, writing, and reading. These skills are used in translating or analyzing verbal problems or to connect the language they know with mathematical terms that may be unfamiliar to them. Reflection is another key component of the mathematics curriculum. When students reflect on their responses and incorrect answers, they gain a new understanding of mathematics.

Speaking the language of mathematics in the classroom enables students to make mathematics a part of their lives and clarify their thinking.

Standard 3: Mathematics as Reasoning

This standard emphasizes reasoning so students can draw logical conclusions about mathematics; use models, known facts, properties, and relationships to explain their thinking; justify their answers and solution processes; use patterns and relationships to analyze mathematical situations; and believe that mathematics makes sense.

A major goal of mathematics instruction is to help students develop the belief that they have control over their own success or failure and they have the power to do mathematics. It is important for students to realize that mathematics is more than just memorizing rules and formulas; rather, it makes sense, it is logical, and it can be fun. Learning to reason and make sense of mathematical concepts happens during problem solving and communicating. Working on projects with other students provides opportunities for all students to express their ideas and to respect and support others' ideas.

Mathematics as reasoning requires critical thinking, informal thinking, conjecturing, and validating, helping students to see how mathematics makes sense. Students should be encouraged to discuss and justify their solutions through the use of manipulatives, physical models, and other concrete objects.

Standard 4: Mathematical Connections

The study of mathematics should include opportunities to make connections so that students can link conceptual and procedural knowledge, relate various representations of concepts or procedures to one another, recognize relationships among different topics in mathematics, use mathematics in other curriculum areas, and use mathematics in their daily lives.

This standard encourages students to make connections between and among mathematical ideas and everyday experiences. Rather than teaching computation, geometry, measurement, and problem solving as isolated components, the connections among them should be made explicit in the mathematics curriculum. If these connections are made clear, lessons flow into each other and extend over a period of time so students have the chance to explore, discuss, and generalize the connections. Through mathematical activities and concepts, students construct bridges between the concrete and the abstract and learn different ways of representing solutions to problems.

Estimation skills are very useful in everyday experiences and develop students' sense of good judgment and logical reasoning to make decisions.

Standard 5: Estimation

The mathematics curriculum should include estimation so that students can explore estimation strategies; recognize when an estimate is appropriate; determine the reasonableness of results; and apply estimation in working with quantities, measurement, computation, and problem solving. Estimation is another aspect of mathematics that interacts with number sense and spatial sense to help students develop insights into procedures and concepts. Understanding estimation skills enables students to deal with everyday quantitative situations, such as estimating the total cost of groceries or estimating the time it takes to walk to a friend's house. These skills develop through experiences which enhance a student's understanding of mathematical inquiry and understanding.

Estimation skills are very useful in everyday experiences and develop students' sense of good judgment and logical reasoning to make decisions. Emphasis on computational estimation helps students develop creative thought processes and fosters a sense of mathematical power.

Standard 6: Number Sense and Numeration

The mathematics curriculum should include whole number concepts and skills so that students can construct number meanings through real-world experiences and the use of physical materials; understand our numeration system by relating counting, grouping, and place-value concepts; develop number sense; and interpret the multiple uses of numbers encountered in the real world.

Students must understand numbers if they are to make sense of the ways numbers are used in the world around them. The numbers are used to measure, identify location, and quantify. When numbers have meaning, for example, phone numbers and ages, students' experiences with numbers are more beneficial. A link between the everyday world and mathematics is developed through exploratory experiences with numbers and number relationships.

Important components of the development of number sense and ideas include counting skills, ordering, rounding, and comparing numbers. Counting forwards and back and skip counting signify advancements in a students' development of number ideas. Another critical component is the understanding of place value and the development of comprehension of number concepts. The concept of place value grows out of grouping experiences.

A developmental approach to computation encourages students to actively be involved in problem solving, discussing their work, raising questions, and validating solutions.

Standard 7: Concepts of Whole Number Operations

Whole number operations include concepts of addition, subtraction, multiplication, and division of whole numbers so that students can develop meaning for the operations by modeling and discussing a rich variety of problem situations; relate the mathematical language and symbolism of operations to problem situations and informal language; recognize that a wide variety of problem structures can be represented by a single operation; and develop operation sense.

This standard emphasizes the understanding of fundamental operations of addition, subtraction, multiplication, and division and applying these operations meaningfully to real-world situations. Instruction on operation sense focuses on concepts and relationships rather than on computation skills. Using mathematical language and intuition, students connect these operations with problem situations.

Standard 8: Whole Number Computation

The mathematics curriculum should develop whole number computation so that students can model, explain, and develop reasonable proficiency with basic facts and algorithms; use a variety of mental computation and estimation techniques; use calculators in appropriate computational situations; and select and use computation techniques appropriate to specific problems and determine whether the results are reasonable.

This standard encourages teaching students a variety of ways to compute, as well as the usefulness of calculators and computers in solving problems with large numbers requiring complex computations. A developmental approach to computation encourages students to actively be involved in problem solving, discussing their work, raising questions, and validating solutions.

Standard 9: Geometry and Spatial Sense

The mathematics curriculum should include two and three-dimensional geometry so that students can describe, model, draw, and classify shapes; investigate and predict the results of combining, subdividing, and changing shapes; develop spatial sense; relate geometric ideas to number and measurement ideas; and recognize and appreciate geometry in their world.

Geometry helps students make sense of the world they live in. Geometry and spatial sense are intriguing and motivating, and tapping into this motivation can improve number understanding and skills and foster an interest in mathematics. Through geometry, students investigate, experiment, and explore with everyday objects.

Collecting, describing, and interpreting data helps students make decisions and predictions with a spirit of exploration and investigation.

Standard 10: Measurement

The mathematics curriculum should include measurement so that students can understand the attributes of length, capacity, weight, mass, area, volume, time, temperature, and angle; develop the process of measuring and concepts related to units of measurement; make and use estimates of measurement; and make and use measurements in problems and everyday situations.

Measurement connects mathematics to everyday experiences and develops mathematical concepts and skills; therefore, it is of central importance to the curriculum. Estimation is emphasized to help students understand the attributes and the process of measuring different items. This standard will give students a firm foundation, enabling them to use any measurement system.

Standard 11: Statistics and Probability

The mathematics curriculum should include experiences with data analysis and probability so that students can collect, organize, and describe data; construct, read, and interpret displays of data; formulate and solve problems that involve collecting and analyzing data; and explore concepts of chance.

The study of statistics and probability highlights the process of solving real-world problems through questioning, searching for relationships, and conjecturing. Collecting, describing, and interpreting data helps students make decisions and predictions with a spirit of exploration and investigation. Statistics allows students to interpret the world around them using numbers, charts, and graphs; it is a tool for solving problems. These problems or everyday situations incorporate measurement, estimation, computation, and problem solving.

Standard 12: Fractions and Decimals

The mathematics curriculum should include fractions and decimals so that students can develop concepts of fractions, mixed numbers, and decimals; develop number sense for fractions and decimals; use models to relate fractions to decimals and to find equivalent fractions; use models to explore operations on fractions and decimals; and apply fractions and decimals to problem situations.

The power of numbers is evident in the usage of fractions and decimals in measurement, probability, and statistics. Understanding fractions and decimals is a sign that a student has a sound understanding of numbers. Instruction of these concepts helps students explore relationships among numbers and builds initial concepts about equivalence and order. Using physical materials and real-world situations are suggested since this is a higher-level concept than understanding numbers and possessing number sense.

Working with and identifying a variety of patterns helps students develop the ability to organize and classify information.

Standard 13: Patterns and Relationships

The mathematics curriculum should include the study of patterns and relationships so that students can recognize, describe, extend, and create a wide variety of patterns; represent and describe mathematical relationships; and explore the use of variables and open sentences to express relationships.

There are many patterns in our world, both in the classroom and outside of it. Working with and identifying a variety of patterns helps students develop the ability to organize and classify information. There are patterns in numbers, shapes, and measurement that encourage students to make connections with the world they live in. A classroom is a rich environment to study patterns and to discover relationships among objects and identify properties within the objects. Color, shape identification, directions, orientation, size, and number relationships are among the many concepts that are involved in pattern recognition. Students should be encouraged to explore and discover different patterns and relationships within their world.

Standard 14: Algebra

The mathematics curriculum should include explorations of algebraic concepts of variable, expression, and equation; represent situations and number patterns with tables, graphs, verbal rules, and equations, and explore the interrelationships of these representations; analyze tables and graphs to identify properties and relationships; develop confidence in solving linear equations using concrete, informal, and formal methods; investigate inequalities and nonlinear equations informally; and apply algebraic methods to solve a variety of real-world and mathematical problems.

This standard applies to the upper grades and bridges the gap between the concrete mathematical concepts and the abstract mathematical concepts. The exploration of algebraic concepts in an informal way builds a foundation for the formal study of algebra in high school. This exploration should emphasize data, graphs, physical models, and other mathematical representations that will help students generalize number patterns and build confidence in their ability to abstract relationships from contextual information. Students use a variety of representations to describe and interpret those relationships. Through the investigation of patterns in number sequences and formulations of verbal rules to describe patterns, students communicate using mathematical terms and concepts.

Standard 15: Number Systems and Number Theory

Students should come to understand and appreciate mathematics as a whole, not as isolated facts and rules.

In the upper grades, the mathematics curriculum should include number systems and number theory so that students can understand the need for numbers beyond the whole numbers; develop and use order relations for whole numbers, fractions, decimals, integers, and rational numbers; extend their understanding of whole number operations to fractions, decimals, integers, and rational numbers; understand how the basic arithmetic operations are related to one another; and develop and apply number theory concepts (e.g., primes, factors, and multiples) in real-world and mathematical problem situations.

Students should come to understand and appreciate mathematics as a whole not as isolated facts and rules. This standard is the underlying structure of mathematics, helping students logically relate knowledge about the world they live in to the whole number system which extends to the rational number system and beyond. Additionally, problem solving requires the knowledge of number theory which can be enjoyable and interesting. Finding the connections between the prime structure of a number and the number of its factors is a challenging activity for students that involves the knowledge of number theory. Having an understanding of number systems and number theory demystifies the world of mathematics.

Concluding Remarks

The curriculum standards discussed in this chapter are incorporated in the three interdisciplinary units in chapter six, "Linking Math to Our World." These standards are value judgments based on a broad perspective of teaching and learning and are a way to meet the overarching goals of the mathematics curriculum. Standards are the first step in the reformation process in school mathematics and contribute to developing mathematically powerful and literate students. The goal is for your students to become problem solvers and mathematicians prepared for the twenty-first century.

Connections

An Integrated Approach to Mathematics

Thinking about mathematics in an integrated learning environment affords teachers and students an opportunity to conceive of math processes and skills as connected to other facets of the curriculum. Mathematics is integrated with other content areas to present to students a curriculum that mirrors the world around them. Students come to view mathematics as part of their lives, as having purpose and meaning to them, rather than as an isolated series of computations and processes where connections are meaningless.

As they participate in an integrated curriculum, their own experiences and knowledge about the world are incorporated into the classroom activities.

An integrated approach views the curriculum as one entity. Through an integrated curriculum approach, teachers and learners construct meaningful connections between ideas and concepts presented in the classroom and the world around them. As they participate in an integrated curriculum, their own experiences and knowledge about the world are incorporated into the classroom activities. With the newly constructed connections, the integrated curriculum approach "represents a way to avoid the fragmented and irrelevant acquisition of isolated facts, transforming knowledge into personally useful tools for learning new information" (Lipson, Valencia, Wixon, & Peters, 1993, p. 252).

Integrating the curriculum is an approach to learning that is not new to the educational community. John Dewey was, perhaps, a leader of this idea of integrating the disciplines to better reflect the world outside the classroom. His thoughts have permeated classrooms over the years, and, consequently, there are many teachers integrating the disciplines, most commonly through thematic units. Dewey believed that by integrating curricular domains and connecting ideas to real-world experiences, students would construct knowledge that was meaningful to understanding the ways in which we live in society. The connections students build in bringing together the various aspects of knowledge result in a web of relationships (Rogoff, 1990). Dewey (1916) talked about the promise held in constructing relationships and connections by saying, "an ideally perfect knowledge would represent such a network of interconnections that any past experience would offer a point of advantage from which to get at the problem presented in a new experience" (p. 340). The network of interconnections is built from students bringing together previous knowledge and new concepts within an integrated setting to construct new interpretations and experiences. As students and teachers engage in an integrated curriculum where concepts are approached from many angles, pathways to understandings become complex and multifaceted.

Through an integrated curriculum approach, students are provided with endless possibilities for mathematical literacy.

Through an integrated curriculum approach, students are provided with endless possibilities for mathematical literacy. There are many pathways to access in constructing meanings about mathematical concepts and processes, and an integrated curriculum approach validates and honors these various pathways. Mathematical skills and concepts are embedded within a theme, creating a context which provides meaning and purpose for the students (Griffiths & Clyne, 1988). As students come to explore the curriculum standards identified earlier in chapter four "Curriculum Standards," connections between the classroom world and the world at large are strengthened. Students come to appreciate the value and importance of mathematical literacy as situations are presented that reflect similar real-world situations in their lives.

Integrating the curricular areas to highlight the connections and relationships between and among processes and skills requires you to consider ways in which integration will be most successful and effective. Merely putting together two or more content areas does not ensure learning will be most effective. It is not enough to have the activities transcend subject matter lines (Brophy & Alleman, 1991). Rather, the connections must be authentic and relevant to be purposeful to the students. "Effective curricular integration occurs when the content from one subject area is used to enhance or enrich the content of another" (Lapp & Flood, 1994, p. 418). So, in thinking about the role of mathematics within integrated learning, it is important to highlight some factors that support the nature of integrated learning and its effectiveness in the classroom context.

Features of Integrated Learning

There are four important features to emphasize as you begin to integrate your instructional program. These include:

- teachers as learners and researchers
- flexibility and balance
- understanding the direction and goals of the curriculum
- creating space for dialogues and conversations

These four features will move the integrated curriculum approach from putting subjects together for the sake of integration to interweaving the curriculum to construct deeper, more meaningful understandings of how we live in this world.

Students and teachers are interacting with each other in a variety of ways and on many different levels.

The first factor, teachers who perceive themselves as learners and researchers, recognizes the importance of continuously searching for more effective and diverse methods of instruction. They know students come to tasks with different perspectives. By honoring the many pathways students take to solve the problem, you are able to learn the thinking patterns of your students and create a more complete picture of them as learners, as mathematicians. The second factor, flexibility and balance, creates opportunities for learning to occur in an authentic context. Knowing that you do not have to stop a discussion or activity because it is time for math relieves pressure on you to make sure the students are receiving the appropriate amount of instruction. With an integrated approach, content areas are balanced throughout the instructional year.

The third and fourth features, knowing the goals and direction of the curriculum and providing opportunities for dialogue in socially negotiated contexts, support an integrated approach to learning. Teachers implementing an integrated curriculum must be knowledgeable about the goals of the curriculum decided for the students because of the diversity in pathways of understandings. This knowledge enables the teacher to plan and prepare for varied possibilities students will bring to the learning event. Along with knowing the goals and direction, a socially situated context where diverse interactional patterns are prevalent is the fourth feature of an integrated approach to the curriculum. Students and teachers are interacting with each other in a variety of ways and on many different levels.

There are more dialogues, discussions, partner reading and sharing times, conversations, and negotiations as students and the teacher come to understand the concepts and ideas presented. With diverse interactional patterns provided in the classroom setting, there are "opportunities for students to explore meaning together, creating and investigating new possibilities for interpretation" (Leal, 1993, p. 117).

With these four features of an integrated approach to learning in mind, where does mathematics instruction fit into the picture? As a teacher committed to providing an integrated curriculum to your students, it is important to recognize the commonalties between mathematical literacy and other learning domains. The commonalties include learning through language and learning in authentic and relevant contexts. Focusing on these similarities will help you to view mathematics as an integrated whole, connected and interrelated to all other content areas.

Language encourages students to clarify, refine, and consolidate their thinking as they explore mathematical problems.

Connections to Other Curriculum Components

Mathematics in an integrated approach to learning is an area of the curriculum where students utilize communication skills to gather and compare data, as well as study patterns and order (Braddon, Hall, & Taylor, 1993). Mathematics is learned through language. Students engage in activities where communicating math concepts through reading, writing, drawing, and speaking are emphasized. "Mathematical language and everyday language are interrelated, as are mathematical thinking and everyday thinking" (California Department of Education, 1992, p. 35). Opportunities to read and explore ideas in word problems, to discuss possible ways of solving the problem, and processing ideas in math journals are all possibilities of how math and language are interconnected and woven together. The relationship between math and language is further supported by the NCTM when one of the overarching goals for mathematics education includes "learning to communicate mathematically" (1989, p. 6). Language encourages students to clarify, refine, and consolidate their thinking as they explore mathematical problems. Students use language to make sense of their worlds and events occurring within the worlds. "It is a natural communication system that we can use to describe our world and communicate our experiences" (Whitin & Wilde, 1992, p. 5).

Mathematics, as with other disciplines, should be presented in authentic and relevant contexts. Without even realizing it, you and your students are engaged in mathematical reasoning and problem solving on a daily basis. Before beginning school, young children have an understanding of addition and subtraction. They can put items together to represent more and take them away to represent less. Students see patterns and shapes all around them in the environment—at home, at the park, in the grocery store, etc. All of these opportunities to explore and manipulate mathematical concepts in authentic, relevant contexts can be mirrored in the classroom setting. Authentic contexts engage students in constructing knowledge that is framed by the culture in which it was created. For example, estimation is one of the most important strategies used in our society. We constantly pose questions to ourselves, such as, "I wonder if there is enough?" "How long will it take?" "How much can be purchased?" etc. As students encounter estimation as a mathematical concept, it should be approached in a relevant manner. Providing opportunities for students to estimate in purposeful ways, such as estimating how long it will take to walk to the playground and back, or how many apples to buy for the fruit salad, or how many pieces of candy will everyone receive from a bag of hard candies, will develop more complete understandings of estimation. Placing mathematics in real life enables students to value the processes in mathematical reasoning and problem solving.

Authentic contexts enable students to realize that problems are rarely solved in one step and that there are many steps and ways to approach the situation.

Authentic and relevant contexts are also discovered in other settings, such as cooking and sports. Recipes have many mathematical concepts, such as fractions, estimations, time, and measurement. Students come to explore and investigate these concepts as they read and understand the recipe and directions. Authentic uses of mathematical concepts and ideas are prevalent in sporting events. Students use number operations, distance, measurement, statistics, and probability when participating in sports. Through relevant and authentic contexts, students discover the importance of math in their everyday lives. They are able to organize their ideas about the world and make sense of what they see.

Authentic contexts for mathematics value the informal knowledge students possess as they enter into the problem. Students often utilize common sense knowledge when solving problems. This intuitive knowledge is unconnected to the formal concepts, procedures, and symbols learned in the classroom setting (Battista & Larson, 1994). Honoring the intuitive knowledge often provides for students diverse possibilities in solving the problem. Authentic contexts enable students to realize that problems are rarely solved in one step and that there are many steps and ways to approach the situation.

Connections to Literature

The relationship between math and other curricular domains is highlighted through language and authentic contexts. As you continue to investigate the nature of integrated learning, one area that may provide a realistic place to begin is integrating mathematics with literature. Literature has proven to be an excellent source of discovering mathematical concepts and processes in real-life contexts and situations. Through literature students are able to break down the often imposed dichotomy between learning mathematics and living mathematics (Whitin & Wilde, 1992). Literature can provide students with a basis to explore the idea that essentially, mathematics is a communication system that is composed of symbols rather than words. The stories students read and listen to have many mathematical concepts embedded within their texts. Some literature selections are explicitly connected to mathematical concepts, such as counting books and books with concepts of time, shape, proportion, measurement, etc. Other literature selections have implicitly embedded mathematical concepts. Math concepts often are not the focus of these stories, but in reading them, mathematics is discovered. Literature selections such as *The Doorbell Rang* (Hutchins, 1986), *Jumping the Broom* (Wright, 1994), and *The Patchwork Quilt* (Flournoy, 1985) provide students with rich, interesting contexts to understand how mathematical concepts are interwoven into the threads of the curriculum. Literature supports the idea that authentic, relevant contexts are engaging for students as they investigate math concepts and strategies. Regardless of whether math is discovered explicitly or implicitly in the texts, "literature can be a starting point for looking at mathematics in a new way—as a tool for solving the problems that are a part of daily life and for considering issues that concern them as citizens" (Whitin & Wilde, 1992, p. 5).

Literature has proven to be an excellent source of discovering mathematical concepts and processes in real-life contexts and situations.

In addition to providing more authentic and relevant contexts to study mathematics, literature also develops the affective and humanistic side of math. For many students, math can be an area of the curriculum where anxiety is high. Approaching mathematical concepts and skills through a literature selection can help to reduce the anxiety by revealing to students that math is all around us and can be approached from many different pathways. The varying genres and styles of text also encourage students to view mathematics from varying perspectives. "Each piece of literature contains a variety of messages to the reader and suggests unique ways of translating those messages into appropriate learning experiences" (Smith, 1995, p. 438). The multiplicity of perspectives can help to deepen and broaden the students' understandings. As students read and respond to literature, there are a number of opportunities to construct connections between their own worlds and the world at large.

Math Standards and Literature

The identified standards in developing mathematical literacy are presented in a variety of literature. Literature can be an impetus to encouraging students to think mathematically. Thinking mathematically involves posing questions, investigating possibilities, drawing logical conclusions, using patterns and relationships, discovering connections, and many other processes. Through the many literature selections, students can explore the different mathematical concepts determined to be valuable for students entering the twenty-first century. The next few pages list the curriculum standards and literature selections that emphasize each standard. There are resources, such as *Math through Children's Literature: Making the NCTM Standards Come Alive* (Braddon, Hall, and Taylor, 1993) and *Read Any Good Math Lately?* (Whitin & Wilde, 1992), that have compiled more thorough lists of texts. This particular list is just a suggestion to begin thinking about the possibilities within literature selections to introduce and emphasize mathematical concepts. In reading over this list, it must be remembered that many concepts overlap with each other. The books are placed in categories that make the most sense, but they are by no means mutually exclusive. In fact, many of the selections have embedded within the story more than one standard or concept. Along with the list of suggested texts, there are a few examples of activity sheets that further support the integration between literature and mathematics.

The next few pages list the curriculum standards and literature selections that emphasize each standard.

◆ **Mathematics as Problem Solving:** (almost any selection will have problem-solving aspects to the story)

Ming Lo Moves the Mountain (Lobel, 1982)

If You Give a Mouse a Cookie (Numeroff, 1985)

◆ **Mathematics as Communication:** (any selection demonstrates to the students that through language, mathematical concepts are being communicated)

The Very Hungry Caterpillar (Carle, 1969)

Melisande (Nesbit, 1989)

Mathematics as Reasoning: (as with problem solving, any selection will encourage students to reason)

Frog and Toad Are Friends: The Lost Button (Lobel, 1970)

The Biggest House in the World (Lionni, 1968)

◆ **Mathematical Connections:** (again, any literature selection will offer the connections between mathematics and other subjects)

Germs Make Me Sick! (Berger, 1985)

Sweet Clara and the Freedom Quilt (Hopkinson, 1993)

◆ **Estimation:**

The Jelly Bean Contest (Darling, 1972)

Moira's Birthday (Munsch, 1987)

In One Day (Parker, 1984)

◆ **Number Sense and Numeration:**

How Many Snails? (Giganti, 1988)

One Watermelon Seed (Lottridge, 1986)

Zero. Is It Something? Is It Nothing? (Zslavsky, 1978)

How to Count Like a Martian (St. John, 1975)

◆ **Concepts of Whole Number Operations and Computations:**

Counting Wildflowers (McMillan, 1986)

So Many Cats (Schenk de Regniers, 1985)

The Doorbell Rang (Hutchins, 1986)

Anno's Mysterious Multiplying Jar (Anno & Anno, 1985)

Emma's Christmas (Trivas, 1988)

The Hilton Hen House (Hinchcliffe, 1987)

◆ **Geometry and Spatial Sense:**

Grandfather Tang's Story (Tompert, 1990)

Flat Stanley (Brown, 1964)

The Boy with Square Eyes (Snape & Snape, 1987)

Rubber Bands, Baseballs, and Doughnuts (Froman, 1972)

◆ **Measurement:**

Pezzettino (Lionni, 1975)

How Big Is a Foot? (Myller, 1962)

8000 Stones (Wolkstein, 1972)

The Line Up Book (Russo, 1986)

◆ **Statistics and Probability:**

Anno's Hat Tricks (Nozaki & Anno, 1985)

Twice Upon a Time (Shapiro, 1973)

Fleet-Footed Florence (Sachs, 1981)

Miss Pickerell and the Weather Satellite (MacGregor & Pantell, 1971)

- **Fractions and Decimals:**

  Alexander, Who Used to Be Rich Last Sunday (Viorst, 1978)

  Dollars and Cents for Harriet (Maestro & Maestro, 1988)

  Henry's Pennies (McNamara, 1972)

  Alice and the Boa Constrictor (Adams & Coudert, 1983)

  Gator Pie (Mathews, 1979)

  A Giraffe and a Half (Silverstein, 1964)

- **Patterns and Relationships:**

  Sam Johnson and the Blue Ribbon Quilt (Ernst, 1983)

  Sea Witches (Robertson & Gal, 1991)

  A Clue in Code (Singer, 1985)

Following this list of suggested literature are two activity sheets enabling you to develop a clearer sense of how to integrate mathematics and literature. The examples are from the stories *Alexander, Who Used to Be Rich Last Sunday* (Viorst, 1978) and *Moira's Birthday* (Munsch, 1987).

All for One Dollar!

What do you think you could buy for a dollar?

With a parent or other family member, go to one or more stores where there would be things for sale for one dollar or less.

Make a list of the things you could buy for one dollar or less. Write the price next to each item.

THINGS I COULD BUY WITH ONE DOLLAR

____

____

____

____

____

____

____

____

____

____

____

Which things on your list would you like to buy?

____

____

____

Which things *would* you buy if you had only one dollar?

____

____

____

____

Total spent, $1.00.

Reprinted from TCM 342 Connecting Math and Literature, *Teacher Created Materials, 1991.*

How Many?

Moira believed that she would need two hundred pizzas and two hundred birthday cakes to feed the two hundred guests she had invited to her birthday party.

Do you think the number of pizzas and cakes she estimated she would need was too much, too little, or just right? Explain your answer.

____

____

____

Now, it's your turn! Estimate how many pizzas and cakes you would need to feed each of these people or groups one slice of pizza and one piece of cake.

Person or Group	Number of Pizzas	Number of Cakes
you		
your family		
your class		
your grade		
your school		
the teachers at your school		
all your relatives		
your town		

Be ready to explain how you decided on the estimates you chose!

Reprinted from TCM 342 Connecting Math and Literature, *Teacher Created Materials, 1991.*

In utilizing literature as a springboard to investigate mathematical relationships and concepts, teachers and students have discovered many possibilities for constructing understandings. Literature is a connecting thread to math and other curricular areas. As your students explore mathematical concepts and ideas in literature selections, they also may discover issues related to science, social studies, and language arts. Students are engaged, interested, and motivated about becoming problem solvers and mathematicians when the context for the math problem is relevant to their lives and connections are made.

Literature is a connecting thread to math and other curricular areas.

Concluding Remarks

An integrated approach for mathematics provides learners and teachers with opportunities to construct meaningful connections between the classroom world and the world at large. The focus of mathematics is on meaning making, problem solving, and discovery. Students are active participants in the classroom, constructing and building connections among ideas and concepts they already know and new ideas and concepts they are being introduced to. Highlighted in the chapter were four features of an integrated approach: teachers continuing to learn and research, balancing the content areas and being flexible with instructional time, understanding the goals of the curriculum, and creating a socially constructed context where learners engage in dialogues and conversations about the concepts presented. When learners and teachers view mathematics as an integrated component of the curriculum, connections and relationships are established. Literature offers students and teachers endless possibilities in constructing connections between mathematics and the real world. Students come to view math occurring in everyday situations that are relevant and purposeful. Integrating math with literature and other curricular domains exemplifies how mathematics is embedded into everyday activities.

Linking Math to Our World

Interdisciplinary Approaches

This chapter offers examples of interdisciplinary units on numbers and number sense, calculating and exchanging money, and sports statistics and probability. These units integrate mathematical concepts and ideas with all other subject areas and everyday experiences. They incorporate the math goals and the curricular standards. Throughout these units, students interact with many mathematical concepts, solve problems, and make connections to the real world. In many ways, students are able to see how mathematics can be utilized in everyday situations.

In many ways, students are able to see how mathematics can be utilized in everyday situations.

Numbers in Our World

Numbers are all around us, on the clock, the television, the VCR, the phone, the calendar, the mailboxes, the cars, the signs, etc. They are extremely important to our world for many reasons. After all, if we did not have numbers, how would we conceptualize our ages, when our birthdays are, or what our addresses and phone numbers are? These are very important numbers and events for young children; hence, this interdisciplinary unit is entitled "Numbers in Our World."

Through various activities, students will communicate, analyze, problem solve, sort, classify, and graph while discovering the multiple uses of numbers. Using word problems, math manipulatives, calculators, rulers, and shapes, students will link conceptual and procedural knowledge and make mathematical connections with their daily lives. This unit focuses on these mathematical connections to everyday experiences, both in and out of school.

Counting the number of snacks is fun as well as work for students newly introduced to numbers.

Children enter school with a perspective that everything is related and there is little difference between work and play. Counting the number of snacks is fun, as well as work for students newly introduced to numbers. Capitalizing on this perspective is imperative. Therefore, this unit on the multiple uses of numbers integrates all subject areas and emphasizes the fact that there are many more activities to do with numbers than just counting and computation. Many literature selections focus on number meanings and numeration. *Moira's Birthday* (Munsch, 1987) is a wonderful example. In this story, Moira wants to have a birthday party and invite all of the children in grades kindergarten through six. And she does so without her parents' knowledge or consent. Through this story, children discover mathematical concepts of numbers, estimation, measurement, money, and problem solving.

In the chapter titled "Connections" is a problem-solving activity that can be done as a whole class or in small groups. You can read the word problems and have the students answer the questions as a group. Then, the students have a chance to pretend it is their own birthday party. They need to estimate how many pizzas and cakes would be needed to feed the family, the class, the grade, etc. The students can write or draw invitations to their birthday party, decide whom they want to invite, and choose what food will be served and the flavor of the cake. Providing the opportunity for students to pretend they are Moira allows them to relate to the story in a different way. After this activity, you can poll the class and graph the flavors of cake and the foods that will be served at the pretend birthday parties. This activity emphasizes the usefulness of a graph and shows how numbers can tell a story.

Another example of literature that incorporates math, science, social studies, and language arts is *How Many Snails?* (Giganti, 1988). In this unique counting book, there are many drawings and classification questions for each page. For example, accompanying a picture of planes, boats, cars, and trucks is the following text:

I went walking to the toy store and I wondered:

- How many trucks are there?
- How many trucks are fire trucks?
- How many trucks are fire trucks and have ladders on them?

This pattern of three questions is followed throughout the story, asking questions about flowers, clouds, foods, animals, and stars. Through these open-ended questions, students discuss their answers and think of new questions. For a whole class activity, students can make their own books with their own unique classification systems. This is a challenge for students to create original sets of questions that match their pictures. While the students are creating their books, they can brainstorm characteristics that are particular to their topics. Using animals as an example, they can discuss physical characteristics, unique behaviors, movement, longevity, diet, and habitat. Or with boats, they can discuss color, size, and speed. Throughout the activity, students use numbers to count how many while they sort and classify the characteristics of their chosen topics.

Throughout the activity, students use numbers to count how many while they sort and classify the characteristics of their chosen topics.

Exchanging Money

In this example, a class is learning about money and whole number computation. An interdisciplinary unit on banking and shopping is set up in the classroom. This unit will teach students a variety of ways to compute, as well as the usefulness of mathematical tools, such as calculators, adding machines, and cash registers in solving problems requiring complex operations. "A developmental approach to computation fosters a problem-solving atmosphere in which children are actively involved in using materials, discussing their work, validating solutions, and raising questions. Placing computation in a problem-solving context motivates students to learn computational skills and serves as an impetus for mastery of paper-and-pencil algorithms" (National Council of Teachers of Mathematics, 1989, p. 45). The students make all of the decisions regarding the bank and the store, and through the hands-on experiences with money and computation, the students can connect mathematics with the real world, specifically, two places they are familiar with—a bank and a store.

The Class Bank. In one corner of the room a table and three chairs has been set up by the students to be the class bank. On the table are some calculators, an adding machine, papers and pencils, deposit and withdrawal slips, and fake money in a money till for the bills and coins. The transactions occurring in this bank include the deposit and withdrawal of money into and from a savings account. The class of thirty students is set up in cooperative learning groups, six groups of five students, and they have decided to rotate the groups so each student has a chance to be either one of the three bankers or one of the two store clerks.

As a class they have discussed the purpose of a bank and the value of money through such literature selections as "Smart" in *Where the Sidewalk Ends* (Silverstein, 1974) and *Alexander, Who Used to Be Rich Last Sunday* (Viorst, 1978). The poem addresses the math concepts of measurement, money, addition, multiplication, and logic. In both selections, a boy cleverly trades his dollar to give him more than he started with. The students discuss exchanging large sums of money for smaller sums of money, based on the value of the coin, not the size. On the following page is an activity that can be used with the poem. This example highlights the concepts of money, addition, multiplication, and logic. Additional activities may include: determining what combination of coins can equal one dollar, discussing how the father really feels about his son's money sense, and pretending to be the father in the poem and explaining the concept of what a dollar is to the son.

The poem addresses the math concepts of measurement, money, addition, multiplication, and logic.

Other literature that can be used includes *Money's Funny Big Book of Math Poems* by Mary Ann Hoberman (1994), *Pigs Will Be Pigs* by Amy Axelrod (1994), and from Silverstein's (1974) *Where the Sidewalk Ends*, "The Googies Are Coming," "Me and My Giant," "For Sale," "One Inch Tall," "Lester," and "Band-Aids."

At the beginning of the unit, each student is given ten dollars in bills and coins that she/he is responsible for. Each group is given five dollars to spend on group items, such as a box of markers which the group is responsible for. (The fake money is supplied by the teacher). One of the learning objectives of the unit is to use the money wisely and not to spend it all, which requires being a problem solver, to think ahead, and think reasonably. The students decide what to do with their own money, whether to save, spend, or a bit of both. However, they are not only responsible for their own money but also for the group money.

Throughout the month students are able to go to the bank to open up a savings account and deposit or withdraw their money. The role of the banker is to monitor the flow of money and guarantee correct amounts. Calculators and adding machines are used to verify their calculations. And in the case of a student spending all of the money too quickly, she/he may take out a loan from the bank and pay it back gradually with interest. Through this activity, students learn about a few of the important services a bank has to offer its customers.

Smart

In this poem a son cleverly trades his dollar to give him more than he started with!

He started with a dollar.

Why did he trade his dollar for two shiny quarters?

____

____

How much money has he lost? ____________

He traded his two shiny quarters for three dimes. Why?

____

____

How much money has he lost now? ____________

He traded his three dimes for four nickels. Why?

____

____

How much money has he lost now? ____________

He traded his four nickels for five pennies? Why?

____

____

How much money has he lost now? ____________

Why does the son think he is smart? ____________

____

Is he smart? ____________ Why? ______________________________

Reprinted from TCM Workshop Notebook: Hands-On Math, *Teacher Created Materials, 1993.*

To incorporate communication skills, students may make a brochure for the bank (modeled after one from a local bank), describing the different services they offer and accounts that customers may open. Also, they may use deposit slips and withdrawal slips designed with a space to write in the amounts of money to be deposited or withdrawn, which emphasizes the concept of place value. Integrating social studies may come in the form of researching the history of banks, banks in the community, and other ways of managing money.

The Class Store. In the other corner of the room, the class has set up a store with an old cash register, paper and pencils, and a calculator. To emphasize the concept of exchanging money, the store will sell school supplies already a part of the classroom, such as books, paper, staplers, scissors, folders, and other school supplies. Students use the money provided at the beginning of the unit to buy their supplies. Bringing in advertisements from local stores enables students to estimate prices for the items in their store. They may design an advertisement for their class store. Students may also create coupons to use at the store. Each student has an opportunity to be store clerk, with the responsibility of adding up the amount the customer owes to the store and giving back the correct amount of change. On the next page, a related activity that can be completed at home is Shopping Spree Scavenger Hunt. Students are to look for different items related to shopping in a store, (i.e., a weekly supermarket flyer, a register tape, a price tag from an article of clothing, etc.). Some follow-up group activities may include locating ten items selling for $.99 per pound, identifying the three most expensive items found in the collection of flyers, and many more. These activities connect mathematical concepts the students are learning in the classroom to real-world experiences.

These activities connect mathematical concepts the students are learning in the classroom to real-world experiences.

Opportunities to visit the store and bank may be limited to encourage students to plan ahead and budget. They will need to calculate how long their supplies will last and how much they should spend. Decisions such as these are related to real life, exposing students to money, computations, and budgeting. Throughout this project, the students visit both the bank and the store and learn the importance of each as a separate entity and as complements to each other.

Shopping Spree Scavenger Hunt

Date Due:____________________________

- ❑ 1. A weekly supermarket flyer
- ❑ 2. Three UPC symbols
- ❑ 3. A register tape from a supermarket
- ❑ 4. A register tape from a department store or pharmacy
- ❑ 5. A price tag from an article of clothing
- ❑ 6. A paper shopping bag

GOOD LUCK!

Student's Name:____________________________

Parent's Signature:____________________________

Shopping Spree Scavenger Hunt

Date Due:____________________________

- ❑ 1. A weekly supermarket flyer
- ❑ 2. Three UPC symbols
- ❑ 3. A register tape from a supermarket
- ❑ 4. A register tape from a department store or pharmacy
- ❑ 5. A price tag from an article of clothing
- ❑ 6. A paper shopping bag

GOOD LUCK!

Student's Name:____________________________

Parent's Signature:____________________________

Reprinted from TCM 199 Math in a Bag, *Teacher Created Materials, 1995*

Additionally, students have made a game board modeled after the Monopoly® Junior game. They have replaced the popular "Park Place" and "St. James Place" with "Main Street" and "Broadway" and other names of local streets. Each student gets a chance to play banker, collecting and giving out money. The game is played during free class time, as well as during recess. The game represents their community and neighborhood, and they have made up the chance cards to relate to their own worlds. For example, instead of saying go to jail, a card says to go to the principal's office. Through this game, the students learn to buy and sell property and railroads and utility companies, collect rent, and negotiate with money and properties. Many mathematical concepts are highlighted in this game, and it encourages students to think and communicate in mathematical terms.

Many mathematical concepts are highlighted in this game, and it encourages students to think and communicate in mathematical terms.

Go, Team!

A month-long interdisciplinary unit on sports has been selected by a class since the local professional football team has recently earned a chance to go to the playoffs and hopes to make it to the Super Bowl. In the next four weeks, students will research and investigate the team and its players, draw conclusions, and predict the team's destiny. Of course, students want the local team to win the Super Bowl, but they must base their predictions on statistics, data collected and analyzed, and then explore the probability of their team winning. "The project encourages students and teachers to use statistical methods to describe, analyze, evaluate, and make decisions" (National Council of Teachers of Mathematics, 1989, p. 70). It promotes mathematical literacy and power by applying mathematics to a real-life situation. Through investigation and exploration, the students will use many other skills, such as reading, writing, drawing, listening, and speaking. The class is set up in cooperative learning groups (six groups of five students). Three groups will research and investigate the offensive team, and the other three groups will research and investigate the defensive team. Each group of students has decided that they will choose a specific player to research and follow in the upcoming games.

The specific information collected falls into three areas:

- about the player
- about the team
- about the sport

The information sheet on the following page should be distributed to all students.

Information Sheet for Interdisciplinary Sports Project

You will need to find out the following information by collecting and analyzing data about a particular sports team and some players. Select three questions to answer from each section of the menu. Keep this information together, perhaps in a portfolio.

I. Investigating the Player

- What is his name? What position does he play? Where is he from?
- Where did he grow up? Where did he go to college? Chart these locations on a map.
- Gather statistics about the player; what is his career standing?
- Does he lead the team in any area (i.e., yards rushing)?
- Figure out percentages for his statistics.

II. Researching the Team

- What is the team's standing in their division? Which division are they in? What are the names of the divisions?
- How many players are on the team? What percentage is in the starting line-up?
- Where is the next game? How far does the team need to travel? Chart the location on a map and measure the distance, using the same map as activity I. What are the predicted weather conditions?
- Which other teams are playing in the playoffs? If eight teams play in the playoffs and two teams make it to the Super Bowl, how many combinations of two teams can there be?
- Investigate the history of the team, such as the city of origin.

III. Analyzing the Sport and the Rules of the Game

- Describe the playing field—the shape, the surface, and the dimensions. How long is a football field? What is the distance from goal line to goal line?
- What is the shape of the football, and why is it shaped like that?
- How does one score?
- What are the scoring patterns? How much is a touchdown worth? a field goal? a safety?
- How many players are on the field? How are they lined up?
- What are the rules of the game (i.e., How many chances are allowed to move the ball 10 yards?)?
- What is the time of play? How long are time outs? half time?

Investigating the Player. Beginning with the player, students find out what position he plays. Different positions warrant different statistics and data. For instance, a quarterback will have different statistics from a wide receiver. Although they both play offense, the quarterback will have a statistic for passing, and the wide receiver will have a statistic for receptions.

Secondly, the students may want to find out where the player is from, both the state and the country. Where did he grow up and go to college? This incorporates geography into the project. The students can chart on a map the hometown of each player and look at how many come from the eastern, western, northern, and southern regions of the country. This information can be obtained in newspaper articles, the media guide supplied by the team's public relations office, a program from a game (if any student has been to a game), or books written about the team and/or the player, such as *Football (1988)* by Norman Barrett or *The Random House Book of Sports Stories* (1990) edited by L. M. Schulman. Another way to gather information is to write to the player or to the team's fan club.

Through this questioning strategy, the students are communicating in mathematical terms while relating to real- life situations that are interesting.

Thirdly, the students find out statistics about the player. They look at career standings and statistics. Again, using the quarterback as an example, his statistics would include: passing yardage, rushing yardage, number of touchdown completions, number of interceptions, number of completed passes, games, or seasons, etc. Some of these statistics will be in whole numbers, so the students must convert them into percentage rates. The students will generate questions to ask each other based on the player's statistics. For example, what percent of the passes thrown in the last game were touchdown passes? Or, what percentage of the quarterback's total yardage was in the air (thrown as passes)? The students can try to stump each other with different kinds of questions. Through this questioning strategy, the students are communicating in mathematical terms while relating to real-life situations that are interesting.

Based on the data collected during the investigation of a player, the students will discuss the player's strengths and weaknesses after analyzing them. They will draw conclusions and make predictions about the performance of their specific player in the next game.

Researching the Team. Although students know what team the player currently plays on, it is interesting to also know if the player ever played for another team and where that former team is located. Other information that is important is the team's standing in its division. A sample of a statistics chart from the sports section of a newspaper is on page 56. This chart can help students to calculate the team ranking in the division (the fact that the team has made it into the playoffs does not necessarily mean it is in first place). Also, the wins and losses are given to help calculate the winning percentage and the team ranking. An important piece of information that may be collected includes how many players are on the team and what percentage of the team is in the starting line-up.

Secondly, the students determine where the next game will be played and how far the team will travel. The students will chart the location on a map and measure the distance the team will travel to the next game, using the same map as the previous geography activity (where the players are from). Also, the weather conditions of each game should be taken into consideration for predictions and review of the game statistics. If it is snowing or raining, the rushing yardage may be considerably less than the passing yardage.

Thirdly, from a sports section of a local newspaper students can find out which other teams are playing in the playoffs and explore the possible combinations of the two teams that will play in the Super Bowl. A sample question may look like this: If eight teams play in the playoffs and only two teams make it to the Super Bowl, how many combinations of two teams can there be? Then, what are the chances of the local team making it to the Super Bowl? The winners of games during round #1 make it to the second round. Then, the winners of the games in round #2 make it to the Super Bowl to compete for the championship.

Sports Statistics

Name ______________________________ Date ______________________

The sports pages of a newspaper often show a variety of sports statistics. One example of the types of statistics shown is the team standings in the National Football League for their particular division rank. Use the chart below to calculate the rankings of the football teams that are listed. The wins and losses are given. First, calculate their percentages and then rank them. The champion with the best record for either division gets #1 rank, and the champion of the other division gets the #2 rank. All other rankings go in order of highest percentage to the lowest percentage. The first one has been done for you.

Remember, to find the percentage, add the wins and losses to get the total number of games. Then divide the number of wins by the total number of games to get your percentage.

Example: 12 + 3 = 15; then, 12/15= .800

NFC

	wins	losses	percentage	rank
San Francisco 49ers	12	3	.800	2nd
Dallas Cowboys	9	6		
New York Giants	14	1		
Green Bay Packers	6	9		

AFC

	wins	losses	percentage	rank
Oakland Raiders	8	7		
Pittsburgh Steelers	11	4		
Indianapolis Colts	13	3		
Buffalo Bills	5	10		

Once the students know which eight teams make it to the playoffs, they can fill out the football playoff schedule which is available in the sports section of the newspaper. Additional information to collect includes historical information about the team and the stadium. Students investigate the history of the team, such as the length of the team's existence, the city of origin, and if it has relocated from another city or state. For the stadium, the group needs to find out the origin of the name and where exactly it is located in the city. This will require a local map. Once they have located it, they can investigate to find out important information, such as the attendance limit. These information-type questions will solicit answers and spark communication using mathematical terms and concepts.

Finally, the students will research the team mascot and why it was chosen. In an example, such as the Kansas City Chiefs, the students can use history books to find out the origin of the mascot. Through other resources, the students can explore what meaning the mascot holds for the team, the players, and the local fans.

These information-type questions will solicit answers and spark communication using mathematical terms and concepts.

Analyzing the Sport and the Rules of the Game. The third component of the project is to gain knowledge about the sport. The students need not become experts; however, it is important for the students to understand some basic concepts of the game which contribute to the uniqueness and excitement of the sport.

First, students can describe the playing field, the shape, the surface, and the dimensions. They may calculate the measurement for the entire length of the field and the distance from goal line to goal line. Another important part of the game includes the markings on the field and the pattern of the markings. Also, knowing how many players are on the field and the formation they set up in is an integral part of the game that follows a pattern, as well. This part of the project can help the students explore patterns and functions to develop mathematical power and help them appreciate the aesthetic beauty of mathematics. Another key element which relates to geometry is the shape of the football, which makes this sport unique. The students need to analyze this further and find out why it is shaped as it is, perhaps investigating the aerodynamics of a football's unique design.

Another important aspect of the game is the scoring system. It is essential to understand the scoring system in order to follow the game. The different ways a team can score need to be investigated, as well as the points allotted to a touchdown, a field goal, a safety, and a conversion. The two-point conversion is when the scoring team decides to run a play which is worth two points as opposed to kicking a field goal which is worth only one point. Knowing how many chances (downs) a team has to move the ball 10 yards is a basic rule of the game, which helps in understanding the game. And some other details can be researched, such as the time of play, the length of time-outs, half time, and the role penalties play in the game.

Assessment and evaluation of the students and their projects is an important element throughout the units. This area will be further addressed in chapter seven "Alternatic Assessment Practices." Possible assessment techniques include portfolios, journal writing, and group and individual projects. Journal writing is a good tool for assessment for any of the units. It offers students an opportunity to review the unit and assess their ideas about the different mathematical concepts that are covered. In their journals, students can write about the struggles and triumphs they experience during the unit, which gives the teacher a window to look through when assessing the student and the overall effectiveness of the unit.

Solving problems, inquiring about solutions, and developing curiosity are what mathematics is all about.

In reviewing the overall unit projects, all subject areas and modes of intelligences are integrated. The unit projects may require the students to read about numbers, banks, and sports teams (reading); locate numbers in the community, find out the history or background information on banks, and locate information on players and teams (social studies); write their own stories related to numbers, create brochures, and write to the player or the team's fan club (writing); classify objects, identify various metals of coins, and record the weather conditions during the games (science); draw, paint, sketch, or trace pictures for books, store advertisements, and the team logo or mascot for a team pennant (art).

Concluding Remarks

All three of these examples of interdisciplinary units offer ideas and activities to integrate mathematics across the curriculum and into the real world. They address the five goals set forth by NCTM: learning to value mathematics, becoming confident in one's own ability, becoming a mathematical problem solver, learning to communicate mathematically, and learning to reason mathematically. These units also address the thirteen curriculum standards for grades kindergarten through eight in exciting and interesting ways. They broaden the realm of mathematics into everyday experiences, which contributes to the power of mathematics. Solving problems, inquiring about solutions, and developing curiosity are what mathematics is all about. Your classroom can provide opportunities for students to become mathematicians and satisfy their curiosity through various pathways to gaining knowledge.

Alternative Assessment Practices

Mathematics and Alternative Assessment

Recall the first scenario in chapter one. The teacher in that scenario was giving a timed math test on computational skills. After the students marked the number incorrect at the top of the page, the teacher collected the papers and recorded them in the grade book. From this evaluation practice, what information is gained about these students as problem solvers, as mathematicians? Is the teacher able to understand the students' thinking processes as problems are solved by looking at a single, one-dimensional score? Are the scores reflective of the students' abilities and interests in mathematics? If your answer is no, timed tests and other chapter-end tests are not the best ways to assess student knowledge and development. What, then, should be utilized? What assessment practices will enable students to demonstrate knowledge and provide teachers with meaningful information about the students' abilities? These concerns are critical as teachers and students begin to question the ways in which mathematics instruction is addressed in the classroom setting. How do the alternative formats for teaching and learning affect the ways teachers assess and evaluate student growth and development?

From this evaluation practice, what information is gained about these students as problem solvers, as mathematicians?

Classroom environments where mathematics instruction occurs within an integrated thematic unit or blocks do not encourage or support end-of-chapter tests or timed tests to evaluate student performance. Students are no longer completing worksheets of computations, looking for single right answers. Rather, math is being taught in a relevant context where problems and solutions mirror those in outside settings. Students are working in cooperative groups, discussions occur on a regular basis, and there are many possible pathways to discovering solutions. With changes in the way the curriculum and instruction are being implemented, there are also changes in the assessment formats. Teachers are now faced with discovering new ways to assess students' growth in mathematical knowledge and evaluate their own abilities to teach math concepts and skills.

The informed decisions teachers make regarding mathematics instruction benefit the students' progress towards math literacy.

As teachers begin to implement alternative assessment practices in evaluating mathematical literacy, it becomes necessary to understand the role assessment plays in the instructional process. Assessment is an integral aspect of the teaching and learning processes. Decisions affecting level of instruction and/or instructional programs are based on results from assessment practices. Teachers and students engage in assessment practices in an effort to determine the best fit for the student and the educational environment (Smith, 1994). This "best fit" reflects the instructional decisions made by the teacher, as well as actions of the students and the classroom environment. It is through assessment that teachers and students come to understand the nature of the learning process, that it is, in fact, a process occurring over time.

Guiding Principles of Alternative Assessment

The adoption of alternative assessment practices to better reflect instructional practices is embraced by three overarching principles defined in the *Curriculum and Evaluation Standards for School Mathematics* (National Council of Teachers of Mathematics, 1989). First, student assessment is integral to instruction. Aligning the mathematics curriculum with assessment creates in students an awareness that what is taught and developed is assessed, and the information gained from the assessment is meaningful and will be utilized to inform future activities and projects (Seely, 1994). The informed decisions teachers make regarding mathematics instruction benefit the students' progress towards math literacy. For example, in working through the curricular standard of fractions and decimals, a teacher may observe and document some students who are having a difficult time conceptualizing the relative size of fractions (i.e., 3/4 is larger than 3/5, or 1/10 is closer to 0 than 2/5 is). Before moving forward and introducing adding and subtracting of fractions, the teacher may provide more opportunities to work with fraction bars and other manipulatives, thereby reviewing the concept of relative size. In doing this, the teacher is aware of students' needs and offers more experiences and instruction to better understand fractions. Utilizing the information obtained from an assessment practice, such as observation, to inform curricular decisions enables the teacher to provide lessons that are appropriate and beneficial as students acquire new understandings.

Aligning assessment with instruction also refers to an emphasis placed on assessing what was taught. If the assessment practice evaluates students' abilities in computation but the curriculum focused on problem-solving and reasoning, the instruction and assessment are misaligned. On the other hand, if students demonstrate their understandings of concepts through open-ended questions and the assessment practice contains many open-ended questions, then curriculum and assessment are considered to be aligned.

Second, there should be multiple ways of assessing students' knowledge and math literacy.

Second, there should be multiple ways of assessing students' knowledge and math literacy. "Because mathematical thinking is complex and has many aspects, the assessment of this thinking requires the use of different sources of information to ascertain students' development in this thinking" (Webb, 1993, p. 2). Assessing students on a variety of tasks and concepts provides a more complete picture of the student as a learner. Students engage in the learning process in different ways. They bring to the learning task their own backgrounds, experiences, and prior knowledge. There will be students who are more successful discovering alternative solutions for open-ended problems; others will perform well in a discussion involving mathematical concepts, and still others will demonstrate understanding on traditional exams. With the multifaceted approaches towards learning and thinking, assessment practices must also be varied. Sound assessment practices expand beyond the one-dimensional paper-and-pencil approach and encompass a multitude of tasks. Assessment should be "committed to sampling a wide range of cognitive processes [and] affective responses" (Valencia, 1990, p. 338).

The third guiding principle envisioned by the NCTM in adopting alternative assessment practices is that there is a convergence of mathematical knowledge. All aspects of this knowledge, including connections among the many ideas, are developed. Students should be provided opportunities to explore mathematical concepts and ideas in a variety of contexts, formats, and problem situations. These opportunities will demand different kinds of mathematical thinking on the part of your students. Utilizing multiple sources encourages students to expand beyond their own knowledge and understandings of mathematical concepts.

With these overarching principles guiding alternative assessment practices in mathematics, the teacher and students are able to view assessment as continuous, dynamic, and, often, process oriented in nature. Alternative assessment practices document and evaluate the ways in which students represent, reorganize, and utilize new information. The focus for alternative assessment is not only on the product but also on the process and performance (Worthen, 1993). Understanding the ways in which students approach and solve problems and tasks provides a foundation in thinking about your students as mathematicians, as problem solvers. Through alternative assessment practices, which highlight the process, students and teachers focus on the gradual, cumulative knowledge built on prior experiences and background knowledge (National Council of Teachers of Mathematics, 1989).

There are a number of alternative assessment practices that have been developed with these guiding principles and process orientation in mind.

There are a number of alternative assessment practices that have been developed with these guiding principles and process orientation in mind. Included, but not limited to, are such practices as portfolios, enhanced multiple-choice assessments, open-ended problems, rubrics, observations, and interviews. Each of these alternative practices contributes to discovering more information about the students' mathematical literacy knowledge. As you and your students begin to implement alternative assessment practices, you might want to keep a few questions in mind.

- Does the student understand the mathematical concepts presented?
- Has the student been engaged in the learning process?
- Is the student making connections between the various ideas and concepts presented in the different contexts?
- What pathways does the student seem to be following?
- What information does the student seem to be seeking?

Beginning with portfolios, these identified assessment practices highlight the many dimensions of mathematical literacy students are able to demonstrate in various contexts and formats.

Math Portfolios

Portfolios have been on the alternative assessment scene for a number of years. They are a collection of student work which is evaluated on rubrics and standards determined by the teacher and, possibly, students. Generally, portfolios are associated with language arts in that many students compile writing portfolios. Recently, however, the concept of a portfolio as an assessment tool has expanded to other disciplines. Mathematics is one content area where constructing a portfolio is a meaningful and effective method of assessing students' thinking and learning patterns. The tasks and problems students complete in math can be easily situated and evaluated within a portfolio. Reflective comments are also included to provide those reviewing the portfolio with some understanding of the thinking processes that were involved in solving the problems and compiling the artifacts.

Portfolios document aspects of mathematical literacy, such as using mathematics to make sense of complex situations, ability to use mathematical processes such as computation in the context of many kinds of problems rather than in isolation, defining and formulating problems, and explaining a mathematical concept either orally or in writing (California Mathematics Council, 1991). In constructing math portfolios, students are able to demonstrate their mathematical thinking processes in a multitude of ways. The math artifacts selected for the portfolio often provide evidence of performance and understanding that go beyond computational knowledge. Included may be such artifacts as written descriptions of different investigations and problems, pictures, graphs and charts, responses to open-ended questions, group reports and solutions, descriptions and diagrams of problem-solving processes, and activities and tasks utilizing a variety of mathematical tools and models, such as manipulatives, calculators, computers, and group projects. The multitude of samples from a variety of sources enables students to demonstrate their strategies and techniques in solving open-ended problems, requiring them to think and reason and not just calculate answers.

The math artifacts selected for the portfolio often provide evidence of performance and understanding that go beyond computational knowledge.

The reflective pieces students include in their math portfolios help to assess growth and achievements in understanding mathematical concepts. Reflections provide students with opportunities to review their own performances, explain the reasons for choosing the processes they used, and identify the next step to be taken (California Mathematics Council, 1991). The reflection process helps students build confidence when they gauge their own performances and observe their growth and development in math. Building math confidence is critical to the success students achieve in learning mathematical concepts. Self-reflections in math portfolios communicate to the student and the teacher the thinking patterns and processes that went into solving the math problems.

Math portfolios provide opportunities to engage students and teachers in the entire process of learning. Students are able to demonstrate their mathematical understandings in a variety of ways for a variety of tasks. Teachers can evaluate a student's knowledge by assessing the whole of the portfolio rather than individual pieces of work.

With the whole of the math portfolio in mind, you can examine work samples to see what your student knows about particular mathematical concepts and note how the student's understanding has developed over time, along with how her/his thinking has been clarified (Ferguson, 1992). You are also able to determine whether the student missed key concepts or did not understand and, therefore, will require additional help. A mathematics portfolio will enable the teacher or other evaluator to try to "get a picture of the student's own thinking rather than whether the student can provide the 'correct' answer that the adult has in mind" (California Mathematics Council, 1991, p. 23). Is the student merely giving memorized responses, or has the student interacted with the ideas and incorporated them into her/his own thinking and knowledge structures? Math portfolios provide the insights necessary to effectively and meaningfully evaluate student performance.

Enhanced multiple-choice assessments provide students with opportunities to solve problems in various ways.

Enhanced Multiple-Choice Assessments

A second alternative assessment practice that offers insights into the thinking processes of students is the enhanced multiple-choice assessment. Unlike traditional multiple-choice questions where students focus on one skill or strategy, enhanced multiple-choice assessments ask students to employ more than one strategy to solve the problem. "They require students to make connections among several concepts to arrive at the correct answer" (Pandey, 1991, p. 16). Enhanced multiple-choice assessments provide students with opportunities to solve problems in various ways. Some problems may be solved algorithmically or by logical thinking, others by trial and error or through a more systematic approach. The pathways students travel to construct understandings about mathematics and the world around them is not dependent upon the format of the question but on the student's own strengths. Enhanced multiple-choice assessments ask students to use several types of conceptual understandings rather than recalling facts or procedures.

Enhanced multiple-choice assessments offer students opportunities to make decisions about how the problems are going to be solved. The steps students take to answer the question may vary. In solving enhanced multiple-choice assessments, students are able to see there are many ways to reach accurate conclusions.

Open-Ended Problems

Another meaningful and relevant assessment practice in developing math literacy in students is the open-ended math problem. Within an open-ended problem, students are presented with a description of the problem situation and a question. Generally, students respond to the prompt in writing. There should be enough openness to encourage more than one solution without being vague or abstract. Open-ended questions have four defining characteristics:

- interesting situations
- enter the problem on many levels
- multiple solutions
- audience

Open-ended problems are developed with interesting situations that are engaging for students and involve many mathematical concepts. Secondly, open-ended math problems allow students to enter at many levels of understanding. Students are able to begin working on the problem with any level of knowledge. The third characteristic of open-ended math problems related to the openness is that there are multiple solutions to the problem. By having multiple solutions to the problem, students come to the task with their own prior knowledge and experiences. And the fourth characteristic is that there is an audience for the work. A need is created for students to learn to effectively communicate mathematical concepts and ideas. The four characteristics guide the open-ended problems.

Open-ended problems are developed with interesting situations that are engaging for students and involve many mathematical concepts.

Rubrics

With the inclusion of multiple pathways to knowledge and understandings, teachers are faced with interpreting the divergent sources of information to construct meaningful evaluative decisions. Taking this information and developing rubrics based on predetermined criteria is another form of alternative assessment. A rubric is an articulated set of criteria set on a scale. Rubrics can be as simple or as complex as you and your students desire. Words and phrases are utilized to generate a "picture" of what behavior or task the student demonstrated or did not demonstrate at each level. The levels can be written as narratives or as short specific phrases. The words and phrases are quite descriptive, making the expectations explicit (Winograd, 1994). There are generally three to six levels of students' understandings. On the following page is an example of a generalized six-point rubric.

Rubrics enable students and teachers to have a clearer, more defined understanding of the ways in which learning and tasks are assessed. Students and teachers are more responsive to the learning process when the expectations are clearly established. Through the process of constructing rubrics, the teacher becomes more cognizant of instructional strategies that are effective and ineffective for students. A better understanding of your students' learning processes emerge as rubrics are developed and actualized in the classroom setting.

Generalized Task Rubric

6 Exemplary Achievement

- Demonstrates deep understanding of major concepts—uses language to describe process or strategy—uses tools, including paper and pencil, calculator, and mental math, effectively and when appropriate—reflects and generalizes about process and purpose

5 Commendable Achievement

- Demonstrates detailed understanding of major concepts—uses language, to a point, to describe process or strategy—uses tools, including paper and pencil, calculator, and mental math, but depends too much on time-consuming paper and pencil computation when the calculator would serve more effectively—reflects and generalizes about process and purpose

4 Adequate Achievement

- Demonstrates a fundamental level of understanding the major concepts—uses language at the literal level—uses tools, including paper and pencil, calculator, and mental math, but depends too much on the calculator when mental math or pencil and paper would serve more effectively

3 Some Evidence of Achievement

- Demonstrates partial understanding of the major concepts—is stronger at doing than at describing with language—solves basic problems at the concrete level only

2 Limited Evidence of Achievement

- Demonstrates a lack of required skills to complete task—attempts task but does not recognize incorrect solutions—hesitates to discuss any aspect of situation

1 Minimal Evidence of Achievement

- Demonstrates a lack of understanding of task—can combine objects to create a set, but makes no connection to symbols or generalized process

0 No participation or response

Reprinted from TCM 778 Math Assessment Grades 5–6, *Teacher Created Materials, 1994*

Observations and Interviews

Alternative assessment practices involving the teacher in a more active role are observations and interviews. Observation is not new to teachers; it is just becoming more systematic and purposeful. Observing students in a variety of contexts in a variety of situations is important to the validity of the assessment (Ryan, 1994). Observations enable you to view your students as they solve mathematical problems and investigate solutions. Observation-based assessment has become a primary source of discovering the thinking patterns of students as they participate in the learning process.

Types of observation notes include anecdotal records and checklists. Both are methods in which the teacher quietly records the behaviors and interactions of the students as they are engaged in tasks and activities. Anecdotal records document students' behavior and actions in a variety of contexts (Seely, 1995). The nature of comments included in the written observations focus on the growth and development of the student and what the student can do rather than what he or she cannot do. Teachers have always watched students and made decisions based on these observations. The difference here is that anecdotal records provide teachers with a systematic process of recording behavior and understandings. Completing anecdotal notes enables you to reflect on the actions and behaviors of individual students or groups of students and to make appropriate decisions. Often the anecdotal records are kept in a portfolio to be accessed at a later date.

Checklists, on the other hand, are ways to record whether a certain mathematical task or concept was addressed.

Checklists, on the other hand, are ways to record whether a certain mathematical task or concept was addressed. They do not judge or assess how well the task was performed, only that it was accomplished. Checklists are based on task analysis (Jasmine, 1993). The range and depth of a checklist is determined by the teacher, depending on the desired purpose for the assessment practice. Checklists provide you with necessary and important information about students' development and knowledge as evidenced by the students themselves. With checklists, teachers are able to scan the identified math concepts and components and determine which students may need additional attention or help, along with which students are ready to proceed in a given direction. The value of checklists is that they focus the observations and are helpful in identifying whether or not students need additional attention in certain areas of the math curriculum.

Another form of alternative assessment involving the teacher is the interview. "Interviews are a method of assessment that allow us to gain insight into students' conceptual knowledge and reasoning during problem-solving" (Huinker, 1993, p. 80). Through teacher conducted interviews, you are able to discover students' thinking and reasoning patterns as they complete math tasks and problems. Before beginning the interview, you want to think carefully about the type of information you are seeking. The purpose will guide the type of questions asked. Asking students to think out loud during the task helps to make salient the ways in which knowledge is being constructed. Probing students' thinking processes as they engage in the task offers insights that are usually hidden in the final product. Interviews, while time consuming, enable teachers to access rich sources of information that are not readily available through other means.

Alternative Assessment and Goals for Mathematics

The assessment strategies encourage divergent thinking and honor the process of discovering possible solutions.

With the descriptions of some highlighted alternative assessment practices at your fingertips, you might be wondering how these formats support the outlined goals and standards for mathematics instruction. Alternative assessment practices enable students to explore and demonstrate mathematical literacy in a number of different ways. The assessment strategies encourage divergent thinking and honor the process of discovering possible solutions. As you and your students develop assessment practices that are aligned with the math curriculum in your classroom and are authentic and relevant, the math standards will be addressed.

Referring to the identified overarching goals for math education outlined in *Curriculum and Evaluation Standards for School Mathematics* (National Council of Teachers of Mathematics, 1989), it is possible to envision how alternative assessment practices validate and honor these goals. The overarching goals for all students, once again, are the following:

- Students learn to value mathematics.
- Students become confident in their abilities to do mathematics.
- Students become mathematical problem solvers.
- Students learn to communicate mathematically.
- Students learn to reason mathematically.

The implications of these goals are that students will have numerous opportunities and experiences in working with mathematical concepts. The alternative assessment practices exploit the multitude of possible pathways learners may travel in constructing meaningful understandings of the concepts.

The first goal, valuing mathematics, is evidenced in the alternative assessment formats of portfolios, anecdotal records, and interviews. Students come to view the role and importance mathematics plays in their own lives as they create their own portfolios and see the investment of the teacher documenting their thinking and learning patterns. Goal number two, becoming confident in their abilities to do mathematics, is supported by all of the alternative practices highlighted in the discussion. In accessing a variety of sources of information, students are no longer tied to demonstrating knowledge in one particular way. There are many ways in which learners demonstrate understandings. Having opportunities to incorporate these possibilities into the math curriculum and assessment will raise the level of confidence in students.

The third goal identified in the document, becoming more competent problem solvers, is addressed as students engage in open-ended problems and enhanced multiple-choice assessments. These alternative assessment practices encourage students to explore divergent thinking patterns and search for multiple solutions. The open-ended problems also support the relevant nature of mathematics in the real world. Problems are generally not solved in one step but take many steps and processes before conclusions can be reached.

The goals established for students as they become mathematically powerful are supported by many alternative assessment practices.

Goals four and five, communicating mathematically and developing reasoning abilities, respectively, are evidenced in portfolios, open-ended questions, observations, and interviews. Through these alternative formats of assessment, students are able to communicate their understandings, in either written or oral forms. The reflective pieces students complete for the portfolios asks them to write about their own mathematical thinking processes. Interviews require that students talk about their thinking patterns. With open-ended questions, students learn to discuss how they solved the problem or tasks. Reasoning skills are embedded throughout the alternative assessment practices. Students demonstrate their abilities to make conjectures, gather evidence, and develop arguments supporting their conclusions.

The goals established for students as they become mathematically powerful are supported by many alternative assessment practices. As you and your students engage in the various practices, it is important to reflect on the goals of the mathematics curriculum and connect the goals to the assessments. With the increased attention on developing assessment practices that take a holistic view of mathematics, teachers and students participate in using multiple methods of assessment and various sources of information to make informed decisions about whether the goals for mathematics are being met.

Concluding Remarks

Alternative assessment practices are exciting possibilities for teachers and students. These alternative practices honor and value the multifaceted nature of learning and teaching. Students come into the learning process through many different pathways, and alternative assessment practices offer students opportunities to demonstrate their knowledge in various ways. There are three guiding principles to consider when engaged in alternative practices for mathematics. These principles include alignment between instruction and assessment, multiple methods of assessment, and multiple sources of information.

Alternative assessment practices enable teachers and students to better understand the complexities of mathematics.

As more and more teachers come to recognize the benefits of alternative assessment, they begin to explore possibilities. Five alternative assessment practices were highlighted in the chapter, portfolios, enhanced multiple-choice assessments, open-ended questions, observations, and interviews. Each of the practices helps the teacher and student to create a more complete and accurate picture of the student as a mathematician, a problem solver. A few examples and suggested ideas were provided as you begin to think about ways to implement alternative assessment practices into your own classroom. There are numerous resources available for teachers interested in learning more about alternative assessment practice as a method of evaluation of student achievement and growth in mathematics. Alternative assessment practices enable teachers and students to better understand the complexities of mathematics. Developing and implementing these practices and others in your classroom will provide you with a glimpse of your students' thinking and learning processes as they investigate and solve mathematical questions.

Emerging Power

The Journey Has Just Begun

Mathematics education is changing in classroom settings throughout the country. Students are accessing knowledge about mathematical concepts and skills in a variety of ways. The pathways students explore as math becomes relevant and meaningful to their lives are diverse and complex. Students construct their understandings of mathematical concepts as they build connections among their own experiences, the classroom, and the world at large. In the previous chapters, you were provided opportunities to build your own connections and understandings. Each chapter offered important insights into mathematics education and instruction in the classroom setting. Perhaps you recognized aspects of your mathematics program in your classroom. Maybe your students regularly engage in meaningful, context-driven problems and situations. Or maybe you have developed an integrated unit, highlighting mathematical concepts and connections to real-world experiences. These, along with many other aspects, are critical in building effective programs where students become mathematically powerful and literate. In addition to recognizing key ideas, it is hoped that new understandings also emerged. Learning more about the constructivist perspective driving the latest reform movement or gaining information on alternative

Students construct their understandings of mathematical concepts as they build connections among their own experiences, the classroom, and the world at large.

assessment practices will support and strengthen your program. Opportunities to rethink your approach and strategies towards mathematics education will enable you to take part in this reform movement.

Recall the scenarios in the first chapter. The first scenario offered a picture of math as it was taught in years past—decontextualized with an emphasis on computation skills, single right answers, and time limits. The second scenario was presented as a possibility in your own classroom—students working together in cooperative groups, discovering multiple solutions and understandings of contextualized problems. The emphasis is on the integrated nature of mathematical processes as connected to everyday experiences. This reflects an empowered mathematics program whereby math becomes a tool for understanding ideas and concepts in the world at large.

The students in your classroom hold the key for the twenty-first century.

Throughout the chapters, an underlying emphasis emerged, that mathematics education is evolving. The NCTM (1989) says, "the best way to bring about reform is to challenge directly the perceptions held by many about the content of mathematics, what is important for students to learn, the job of teaching, what constitutes the work of students, and the professional roles and responsibilities of teachers and administrators" (p. 255). *Math Standards in Action* is a step in the journey towards NCTM's vision of mathematics education in the classroom. As you begin your own journey in facilitating the development of mathematical power and literacy in your students, an abundance of possibilities unfold.

The students in your classroom hold the key for the twenty-first century. They will utilize mathematical ideas and concepts in unique and differing ways. We can only speculate on the advancements to be made in technology and science. The challenge, then, is to provide opportunities for your students to develop the mathematical skills and concepts necessary to compete in this changing world. We hope you accept this challenge and assume responsibility for improving the teaching and learning of mathematics in our classrooms.

References

Adams, L., & Coudert, A. (1983). Alice and the boa constrictor. Boston: Houghton Mifflin.

Anno, M., & Anno, M. (1985). Anno's mysterious multiplying jar. New York: Philomel.

Axelrod, A. (1994). Pigs will be pigs. New York: Four Winds Press.

Barrett, N. (1988). Football. New York: Watts.

Battista, M., & Larson, C. (1994). The role of JRME in advancing learning and teaching elementary school mathematics. Teaching Children Mathematics, 1, 178–181.

Berger, M. (1985). Germs make me sick! New York: Crowell.

Braddon, K., Hall, N., & Taylor, D. (1993). Math through children's literature: Making the NCTM standards come alive. Englewood, CO: Teachers' Ideas Press.

Brophy, J., & Alleman, J. (1991). A caveat: Curriculum integration isn't always a good idea. Educational Leadership, 49(2), 66.

Brown, J. (1964). Flat Stanley. New York: Harper & Row.

California Department of Education. (1992). Mathematics framework for California public schools. Sacramento, CA: Author.

California Mathematics Council. (1991). Assessment alternatives in mathematics. Sacramento, CA: California State Department of Education.

Carle, E. (1969). The very hungry caterpillar. New York: Putnam.

Darling, K. (1972). The jelly bean contest. Champaign, IL: Garrard.

Dewey, J. (1916). Democracy and education. New York: Macmillan.

Dewey, J. (1938). Experience and education. New York: Collier.

Dyson, A. H. (1995). Writing children: Reinventing the development of childhood literacy. Written Communication, 12(1), 4–46.

Ernst, L. (1983). Sam Johnson and the blue ribbon quilt. New York: Lothrop, Lee & Shepard.

Ferguson, S. (1992). Zeroing in on math abilities. Learning, 21(3), 38–41.

Flournoy, V. (1985). The patchwork quilt. New York: Dial Books.

Froman, R. (1972). Rubber bands, baseballs, and doughnuts. New York: Thomas Y. Crowell.

Giganti, P. (1988). How many snails? New York: Greenwillow Books.

Griffiths, R., & Clyne, M. (1988). Books you can count on. Portsmouth, NH: Heinemann.

Grisham, D. L., & Molinelli, P. M. (1995) Cooperative learning. Westminster, CA: Teacher Created Materials.

Hinchcliffe, J. (1987). The Hilton hen house. New York: Ashton Scholastic.

Hoberman, M. A. (1994). Money's funny big book of math poems. Chatham, NJ: Steck-Vaughn.

Hopkinson, D. (1993). Sweet Clara and the freedom quilt. New York: Alfred Knopf.

Huinker, D. (1993). Interviews: A window into students' conceptual knowledge of the operations. In N. Webb & A. Coxford (Eds.), Assessment in the mathematics classroom (pp. 80–86). Reston, VA: National Council of Teachers of Mathematics.

Hutchins, P. (1986). The doorbell rang. New York: Greenwillow Books.

Jasmine, J. (1993). Portfolios and other assessments. Westminster, CA: Teacher Created Materials.

Jasmine, G., & Jasmine, J. (1995). Cooperative learning activities for math. Westminster, CA: Teacher Created Materials.

Johnson, D. W., Maruyama, G., Johnson, R., Nelson, D., & Skon, L. (1981). Effects of cooperative, competitive, and individualistic goal structures on achievement: A metacognitive analysis. Psychological Bulletin, 89(2), 47–62.

Karlsson, M. R. (1996). Motivating at-risk learners. Westminster, CA: Teacher Created Materials.

Lapp, D., & Flood, J. (1994). Integrating the curriculum: First steps. The Reading Teacher, 47, 416–419.

Leal, D. (1993). The power of literary peer-group discussions: How children collaboratively negotiate meaning. The Reading Teacher, 47, 114–121.

Lionni, L. (1968). The biggest house in the world. New York: Pantheon.

Lionni, L. (1975). Pezzettino. New York: Pantheon.

Lipson, M., Valencia, S., Wixon, K., & Peters, C. (1993). Integration and thematic teaching: Integration to improve teaching and learning. Language Arts, 70, 252–263.

Lobel, A. (1970). Frog and toad are friends: The lost button. New York: Harper and Row.

Lobel, A. (1982). Ming Lo moves the mountain. New York: Scholastic.

Lottridge, C. (1986). One watermelon seed. Toronto: Oxford University Press.

MacGregor, E., & Pantell, D. (1971). Miss Pickerell and the weather satellite. New York: McGraw-Hill.

McMillan, B. (1986). Counting wildflowers. New York: Lothrop, Lee & Shepard.

McNamara, L. (1972). Henry's pennies. New York: Franklin Watts.

Maestro, B., & Maestro, G. (1988). Dollars and cents for Harriet. New York: Crown.

Mathews, L. (1979). Gator pie. New York: Dodd, Mead.

Munsch, R. (1987). Moira's birthday. Toronto: Annick Press.

Myller, R. (1962). How big is a foot? New York: Atheneum.

National Council of Teachers of Mathematics. (1989). Curriculum and evaluation standards for school mathematics. Reston, VA: Author.

Nesbit, E. (1989). Melisande. San Diego, CA: Harcourt Brace Jovanovich.

Nozaki, A., & Anno, A. (1985). Anno's hat tricks. New York: Philomel.

Numeroff, L. J. (1985). If you give a mouse a cookie. New York: Scholastic.

Pandey, T. (1991). A sampler of mathematics assessment. Sacramento, CA: California State Department of Education.

Parker, T. (1984). In one day. Boston: Houghton Mifflin.

Piaget, J., & Inhelder, B. (1969). The psychology of the child. New York: Basic Books.

Resnick, L. (1987). Education and learning to think. Washington, DC: National Academy Press.

Robertson, J., & Gal, L. (1991). Sea witches. New York: Dial Books.

Rogoff, B. (1990). Apprenticeship in thinking. New York: Oxford University Press.

Russo, M. (1986). The line up book. New York: Greenwillow.

Ryan, C. D. (1994). Authentic assessment. Westminster, CA: Teacher Created Materials.

Sachs, M. (1981). Fleet-footed Florence. Garden City, NY: Doubleday.

Schenk de Regniers, B. (1985). So many cats. New York: Clarion.

Schulman, L. M. (Ed.). (1990). The Random House book of sports stories. New York: Random House.

Seely, A. (1994). Portfolio assessment. Westminster, CA: Teacher Created Materials.

Seely, A. (1995). Integrated thematic units. Westminster, CA: Teacher Created Materials.

Shapiro, I. (1973). Twice upon a time. New York: Charles Scribner's Sons.
Silverstein, S. (1964). A giraffe and a half. New York: Harper & Row.
Silverstein, S. (1974). Where the sidewalk ends. New York: Harper & Row.
Singer, M. (1985). A clue in code. New York: Harper & Row.
Smith, J. (1994). Standardized testing versus authentic assessment: Godzilla meets Winnie the Pooh. In L. Morrow, J. Smith, & L. S. Wilkinson (Eds.), Integrated language arts: Controversy to consensus (pp. 215–229). Boston, MA: Allyn and Bacon.
Smith, J. (1995). Threading mathematics into social studies. Teaching Children Mathematics, 1, 438–444.
Snape, J., & Snape, C. (1987). The boy with square eyes. New York: Prentice-Hall Books.
Spaulding, C. L. (1992). The motivation to read and write. In J. W. Irwin & M. A. Doyle (Eds.), Reading/writing connections: Learning from research (pp. 177–203). Newark, DE: International Reading Association.
St. John, G. (1975). How to count like a Martian. New York: Henry Z. Walck.
Szetela, W. (1993). Facilitating communication for assessing critical thinking in problem solving. In N. Webb & A. Coxford (Eds.), Assessment in the mathematics classroom (pp. 143–151). Reston, VA: National Council of Teachers of Mathematics.
Tompert, A. (1990). Grandfather Tang's story. New York: Crown.
Trivas, I. (1988). Emma's Christmas. New York: Orchard.
Turner, J., & Paris, S. (1995). How literacy tasks influence children's motivation for literacy. The Reading Teacher, 48, 662–673.
Valencia, S. (1990). A portfolio approach to classroom reading assessment: The whys, whats, and hows. The Reading Teacher, 43, 338–340.
Viorst, J. (1978). Alexander, who used to be rich last Sunday. New York: Atheneum.
Vygotsky, L. (1978). Mind in society. Cambridge, MA: Harvard University Press.
Webb, N. (1993). Assessment for the mathematics classroom. In N. Webb & A. Coxford (Eds.), Assessment in the mathematics classroom (pp. 1–6). Reston, VA: National Council of Teachers of Mathematics.
Whitin, D., & Wilde, S. (1992). Read any good math lately? Portsmouth, NH: Heinemann.
Winograd, P. (1994). Developing alternative assessments: Six problems worth solving. The Reading Teacher, 47, 420–422.
Wolkstein, D. (1972). 8000 stones. Garden City, NY: Doubleday.
Worthen, B. (1993). Critical issues that will determine the future of alternative assessment. Phi Delta Kappan, 74, 444–454.
Wright, C. (1994). Jumping the broom. New York: Holiday House.
Zslavsky, C. (1978). Zero. Is it something? Is it nothing? New York: Franklin Watts.

Teacher Created Materials Resource List

TCM 068 Graph Art Puzzles

TCM 093 Holiday Graph Art

TCM 095 Simple Graph Art

TCM 096 Challenging Graph Art

TCM 185 Using Math Manipulatives: Money

TCM 186 Using Math Manipulatives: Base Ten

TCM 187 Using Math Manipulatives: Calculator

TCM 188 Using Math Manipulatives: Bear Counters

TCM 199 Math in a Bag (Primary)

TCM 200 Early Childhood Units for Math

TCM 237 Money Thematic Unit (Intermediate)

TCM 342 Connecting Math and Literature (Primary)

TCM 605 Interdisciplinary Units: Heroes (Challenging)

TCM 656 Cooperative Learning Activities for Math (Primary)

TCM 657 Cooperative Learning Activities for Math (Intermediate)

TCM 770 Math Assessment Grades 1–2

TCM 774 Math Assessment Grades 3–4

TCM 778 Math Assessment Grades 5–6

TCM 790 Teaching Basic Skills Through Literature: Math

TCM Workshop Notebook: Hands-On Math